Review actions for more info

PLANNING FOR URBAN QUALITY

The issue of urban design can often be controversial, with many expressions of discontent and unease with the form in which new buildings or changes to existing buildings take place.

Planning for Urban Quality examines the achievement of quality in the urban environment, within a planning context. Tracing urban design from its roots, the authors draw on both historical and current practices to examine the key physical, political and economic forces at play and the social pressures and impacts brought about by both failures and achievements in urban design. It moves from a historical review to providing topical examples of public controversy with a commentary on modern urban design and a look ahead at the future of planning and development.

This highly illustrated critique of towns and cities draws on examples from across Western Europe, South Africa and the USA to examine both public and private sector development practices, controls and fiscal policies within a diverse range of localities. The authors indicate the need for a reinstitution of regional–provincial approaches, for closer co-ordination between sectors, and for revised fiscal policies in planning and development so that the potential for improvement is realised and the quality of urban social experience and environments can be enhanced.

The achievement of real and lasting quality in our urban environments has never been more important and controversial than it is now. This book provides a deeper understanding of the many diverse strands of urban quality by examining how it can be achieved and, conversely, what it can achieve for us by way of improving and shaping our lives. The authors provide a firm basis from which to analyse both current and historical urban planning achievements and to assess the relevance and value of urbanscapes.

Michael Parfect is an architect planner, formerly Borough Planning Officer for a Surrey District Council. **Gordon Power** is a consulting architect in private practice at Bristol and has been a lecturer at Bristol and Bath universities. The authors both trained as architects at the Bartlett School, University College London, and their combined professional experience forms the basis of this book.

PLANNING FOR URBAN QUALITY

Urban Design in Towns and Cities

Michael Parfect and Gordon Power

with

LDR International, Inc.

Routledge

London and New York

First published 1997
by Routledge
11 New Fetter Lane, London, EC4P 4EE

Simultaneously published in the USA and Canada
by Routledge
29 West 35th Street, New York, NY 10001

© 1997 Michael Parfect and Gordon Power

Typeset in Garamond by
J&L Composition Ltd, Filey, North Yorkshire

Printed and bound in Great Britain by
Butler & Tanner Ltd, Frome and London

All rights reserved. No part of this book may be reprinted or reproduced or utilised in any form or by any electronic, mechanical, or other means, now known or hereafter invented, including photocopying and recording, or in any information storage or retrieval system, without permission in writing from the publishers.

British Library Cataloguing in Publication Data
A catalogue record for this book is available from the British Library

Library of Congress Cataloging in Publication Data
Parfect, Michael, 1931–
Planning for urban quality: urban design in towns and cities/ Michael Parfect and Gordon Power.
p. cm.
Includes bibliographical references and index.
1. City planning. 2. Cities and towns. 3. Architectural design. 4. Quality of life. I. Power, Gordon, 1932–
II. Title.
HT166.P66 1997
307.1'216–dc20 96–41631

ISBN 0–415–15967–9
ISBN 0–415–15968–7 (pbk)

To Maurice Kinch,
who loved quality and
truth in all things

CONTENTS

List of figures — xii
Notes on authors — xiv
Acknowledgements — xvi
List of abbreviations — xvii

PREFACE
Purpose and structure of book — 1
The role of design guidance — 1
Design literature — 2

PART I: INTRODUCTION
HISTORICAL REVIEW/SETTING THE SCENE — 7
Society and environment in disarray: the tyranny of the car — 7
Historical perspective: background to urban quality in Britain — 12
THE CASE FOR URBAN QUALITY — 19
The need for promotion via local government — 19
The planning situation today — 21
Design guidance — 22
PART I: SUMMARY/KEY ASPECTS — 25
Historical review/setting the scene — 25
The role of design guidance in pursuit of urban quality — 25

PART II: TOWN PLANNING FRAMEWORK
CONTEXT AND MACHINERY FOR DESIGN ADVICE — 29
Resources and scope — 29
New versus old — 30
THE STATUTORY PLANNING FRAMEWORK: POLICY GUIDANCE — 32
Environmental principles — 32
Economic activities — 33
The historic environment — 34
Transport and communications — 34
DEVELOPMENT PLANS — 34

Local plans	35
Other plans	35
Development and planning briefs	37
DEVELOPMENT CONTROL	39
CONSERVATION AND PRESERVATION	42
Origins, concepts and differences of purpose	42
Powers and procedures of local authorities	43
Protection and enhancement: resources and practicalities	45
Public attitudes towards development and conservation	45
Benefits of conservation in practice	46
Enhancement opportunities	48
Unlisted buildings in conservation areas: the need for greater control	49
The way ahead	50
THE DEVELOPMENT PROCESS IN CONTEXT	51
Local objections/representations	54
Committee presentation/information	55
Appeals	55
Legal agreements/'planning gain'	56
FURTHER MEASURES FOR GUIDANCE AND CONTROL	56
Public consultation and participation	56
Sensitive areas/complex or major schemes	58
Developers, architects and the public	59
Planning and design guidance material	60
Land use/floorspace	62
High building strategies/building height policies	64
'NIMBY' AND *LAISSEZ-FAIRE*	69
OTHER MEASURES FOR OBTAINING URBAN QUALITY	70
Traffic measures and parking provision	70
Development phasing control	74
Control of demolition	75
Control of advertisements and signs	77
Nuisances	78
Enforcement of planning control, including deterrence of unauthorised development	80
URBAN ENHANCEMENT MEASURES AND PROVISIONS	83
Galbraith: 'private affluence and public squalor' dictum	83
Private and public sector joint action potential	83
Property/land value enhancement	85
Preservation pays	87
Environmental improvement via planning gain; 'percent for art'; street schemes; grant aiding; and pedestrianisation schemes	88
IN PURSUIT OF URBAN QUALITY: PUBLICITY, CONTROVERSY AND EDUCATION	98
Introduction	98
Influence of the press and the media	99
Publications, publicity and pronouncements	100

Major architectural controversies	101
New buildings in historic contexts	103
ACHIEVING QUALITY IN URBAN DESIGN	104
The Prince of Wales's 10 principles of (traditional) design	105
Francis Tibbalds: urban design guidelines/personal appreciation	110
Other influential publications	113
Design control: the RIBA/RTPI Design Accord	115
Other initiatives: The DoE's quality campaign	118
EDUCATION FOR URBAN QUALITY	119
Design awareness of the lay public	119
Design education	121
Environmental education	123
PART II: SUMMARY/KEY ASPECTS	125
Context for design advice	125
Conservation and preservation	125
The development process in context	126
Further measures for guidance and control	126
Urban enhancement measures and provisions	127
In pursuit of urban quality: publicity, controversy and education	127

PART III: PERSPECTIVE ON URBAN QUALITY

A WAY FORWARD: INVESTMENT IN URBAN QUALITY	133
New directions	134
Urban quality: the embodiment of meaning	135
PHILOSOPHY OF CONTEXT	137
DEVELOPMENT QUALITY: COMPONENT FACTORS	145
Finance and maintenance	146
Amenity	146
Traffic, parking and access	146
Security and safety	147
Order and organisation	147
THE PUBLIC REALM: NEGLECT, DECAY AND REVITALISATION	148
THE PUBLIC REALM: A HISTORIC BASIS FOR MODERN URBAN DESIGN/QUALITY	152
Introduction and review	152
Urban design in modern times	154
The roots of modern urban design	155
Urban design as 'planned evolution'	156
New towns and cities	158
URBAN QUALITY: SITUATION ANALYSIS	159
Review	159
URBAN SITUATIONS: RELEVANCE AND CAPACITY FOR CHANGE	159
Environmental capacity	163
ISSUES AND INDICATORS FOR URBAN QUALITY	168

CONTENTS

City of Chester: (some main) issues and indicators	168
URBAN SITUATIONS: CASEBOOK	169
Context/capacity	169
Context/conservation	172
Development quality: award winners and other successes	173
Public realm: public/private sectors	174
Public realm: Midlands/North of England/Scotland	174
Public realm: waterfronts	179
The new South Africa: Cape Town	179
South-West Africa: Windhoek, Namibia	189
Design and technology	190
MODERN ARCHITECTURE: DESIGN AND TECHNOLOGY	195
The design process: art, science or market exercise?	195
'The Great Divide': old v. new	198
The case for innovation and progress	200
Into Europe and the 21st century	201
INDICATORS, GENERATORS AND CRITERIA FOR URBAN QUALITY	203
Assessment of situation: a general approach/methodology	203
Design guidance and civic design	204
Public and private sector co-operation	206
SUSTAINABILITY IN URBAN QUALITY	212
Local Agenda 21	212
URBAN QUALITY: OLD AND NEW NEEDS	213
The roots of sustainability and their relevance today for urban quality	213
Sustainable revisions in land acquisition and construction realisation	214
Status and deficiencies of the UK construction industry	216
The husbandry of land	217
MECHANISMS FOR URBAN QUALITY IN A SUSTAINABLE FUTURE	218
The Urban Design Group and its aims	218
Community Development Trusts	219
NOTEBOOK: SPECIAL CONSIDERATIONS	220
Provision for women in the environment	221
Racial, ethnic and religious groups	222
THE WAY AHEAD IN PLANNING AND DEVELOPMENT	222
PART III: SUMMARY OF KEY ASPECTS	225
A way forward: investment in urban quality	225
Philosophy of context	225
Development quality: component factors	225
The public realm: neglect, decay and revitalisation	226
The public realm: a historic basis for modern urban design/quality	226
Urban quality: case studies	226
Modern architecture: design and technology	227
Into Europe and the 21st century	227
Indicators, generators and criteria for urban quality	228

Sustainability in urban quality   228
The way ahead in planning and development   228

Notes   229
Bibliography and References   235
Index   238

FIGURES

1	Extract from *Counter-Attack*: urban design versus 'subtopia'	8
2	Dublin street: visual qualities	11
3	Medieval towns: Lincoln	13
4	Baroque city: Rome	15
5	19th-century city: Barcelona	16
6	20th-century motorcity: London	17
7	The city experienced from the vehicle	18
8	Preservation or conservation? Bristol	43
9	Bristol/Lincoln: Examples of preservation	47
10	Lewes/Brighton: *rus* and *urbs*	65
11	Amex Building, Brighton: context and quality	67
12	Traffic impact and relief: Epsom	71
13	Traffic calming in Lewes	72
14	London, Broadgate/Norwich, Castle Mall: award winners	94
15	Traffic segregation: Bristol shopping arcades	97
16	Bristol: historic and maritime legacy	139
17	Edinburgh: New Town and Castle/Old Town	142
18	Metropolitan/macro scale: London/Fort Worth, Texas	162
19	Local/humanist scale: Greenville, S. Carolina	164
20	Chester: historic city context/capacity	167
21	Bristol: RC Cathedral/Clifton Club	167
22	La Rochelle: historic inner port	171
23	Birmingham: Victoria Square	176
24	Birmingham: Town Hall	178
25	Manchester: GMEX Centre	180
26	Manchester: Light Railway	182
27	Manchester: Civic Scale	183
28	Glasgow: Historical context, new life	183
29	North of England waterfronts: Newcastle/Liverpool	184
30	Waterfronts: bridges and buildings	186
31	Cape Town: context and scale	187
32	Cape Town: Waterfront vitality and success	188
33	Windhoek, Namibia: A City in Transition	190

34 Danish building excellence: recent examples 192
35 Hôtel du Département, Marseilles: modern design and technology 194
36 New technological applications in planning 196
37 Cedar Rapids, Iowa: enhancement of a US regional city 206
38 Southfield, Michigan: Central Area reorganisation for the pedestrian 209
39 Lakeland, Florida: lakeside potential maximised for community 211
40 City centre public transport, Strasbourg 215
41 In conclusion: towns and people: Bristol/Sarlat 223

NOTES ON AUTHORS

Michael A.B. Parfect is an architect/planner with many years' experience in Town Planning, mostly in the public sector, in a variety of locations and situations. He trained in both Architecture and Planning in London between 1950 and 1960. His enthusiasm for the role of urban design and conservation stems fundamentally from his early experience in working with Professor Sir William Holford in his architectural and planning consultancy, and with Alfred Wood, the City Planning Officer for Norwich. He also held civic design and planning posts with the London and Hampshire County Councils, and was Deputy Planning Officer for the GLC expanding town of Swindon in the early 1970s where he dealt with major town centre, housing and industrial development.

For 14 years (1976–90) he was the Borough Planning Officer of Epsom and Ewell, Surrey. He has directly promoted much urban improvement and protection during his professional career. His wide experience in local government places him in a very good position to write in a realistic and informative way on the enhancement of urban quality as practised from within the statutory system of Town Planning, especially in respect of the historic environment.

He retired from local government in 1990, but soon decided to put his experience to wider use by writing about planning and urban design, particularly for the benefit of the non-architect layman. He is an unashamed conservationist and traditionalist, but is equally aware of the need to reconcile modern buildings satisfactorily with the environment. He recognises the place of the layman and of amenity societies, as well as local councils, in helping to address the problems of new development and urban change, and has lectured to various student and amenity groups on planning and architecture.

Michael Parfect has an honours degree in Architecture and a diploma in Town Planning, both from the University of London. He is a Member of the Royal Institute of British Architects, a Fellow of the Royal Town Planning Institute, and a Fellow of the Royal Society of Arts. He has also served since retirement as a District and Town Councillor for Lewes, in East Sussex.

Gordon C. Power is a private sector architect of wide and varied experience, and was for many years the principal partner of a well-known architectural practice based in Bristol and Winchester. He trained at the Bartlett School of Architecture in London in the 1950s, contemporaneously with Michael Parfect, where they both completed the 5-year degree course in Architecture with honours.

Gordon Power and his partnership have carried out many commercial, industrial and residential schemes both in this country and abroad. He has a wide knowledge of the practical and financial aspects of most types of development, as well as being a talented designer. With this background, he has first-hand experience and understanding of the problems of designing modern development in traditional settings, and of implementing many different types of scheme within the constraints of the planning system in Britain and elsewhere. He relinquished his partnership/practice in 1990, but still acts as a consultant and adviser for a range of new developments. He also has wide experience in teaching Architectural Design and related aspects, and as an examiner in the Royal West of England School of Architecture. He has been a lecturer at Bristol and Bath universities.

Gordon Power has an honours degree in Architecture from the University of London. He is a Fellow of the Royal Institute of British Architects, and a Fellow of the Royal Society of Arts, of which he became Chairman of the Western Region in 1990.

ACKNOWLEDGEMENTS

The authors gratefully acknowledge the assistance of Kim N. Way of LDR International Limited (land, design and research consultants) for his general support and advice in the formulation of the text and for contributing several of the visual examples and design schemes reproduced in these pages. Thanks are also due to (amongst many others) Michael Ray (Director of Planning, Hove Borough Council), John Silvester (Director of Planning, Surrey Heath Borough Council), Jim Redwood (Director of Planning, Lewes District Council) and David Oliver (District Architect, West Dorset District Council) for their various contributions, direct or indirect. The list would not be complete without mention of Tristan Palmer and Matthew Smith of Routledge, whose infinite patience has enabled this book to happen, and to Tim Blackman of Oxford Brookes University for his generous and helpful reviewing.

Additional warm thanks are due to Linda Bacon for her unstinting provision of secretarial services over the total period of production and crystallisation of the text, to John and Marion Stewart for their practical assistance and interest, and finally to our wives Christine and Joan for their support and general forbearance.

MABP and GCP

ABBREVIATIONS

AGLV	Area of Great Landscape Value
BTA	British Tourist Authority
CCT	Compulsive Competitive Tendering
CDA	Comprehensive Development Area
CPD	Continuing Professional Development
CPRE	Council for the Protection of Rural England
DoE	Department of the Environment
DpT	Department of Transport
EC	European Community
EU	European Union
FSI	Floor Space Index
HGV	heavy goods vehicle
LBC	Listed Building Consent
LPA	local planning authority
NIMBY	'Not In My Backyard'
PAG	Planning Advisory Group
pd	permitted development
POA	Primary Office Area
PPG	Planning Policy Guidance Note
PSA	Property Services Agency
RFAC	Royal Fine Art Commission
RIBA	Royal Institute of British Architects
RTPI	Royal Town Planning Institute
SPAB	Society for the Protection of Ancient Buildings
SPZ	Simplified Planning Zone
SRB	Single Regeneration Budget
TPO	Tree Preservation Order
TPP	Transport Preparation Pool
TPU	Transportation Planning Unit
UDG	Urban Design Group
UDP	Unitary Development Plan

PREFACE

PURPOSE AND STRUCTURE OF BOOK

We live in a period of modern history when, for any thinking person, the achievement of real and lasting quality in our urban environments has never been more important – but at the same time, never more controversial, elusive and problematic.

There is a need, *inter alia*, to take stock of the machinery we have available for achieving urban quality, and to see how much of it is still useful and valid in a rapidly changing world, adapting, amending or replacing it as necessary. More fundamentally, however, we need to define what we mean by urban quality. We have to see, in other words, how urban quality is (or can be) achieved, and conversely what it can achieve for us by way of improving and shaping our lives. To do this, we must look continually below the surface, at all levels, to appreciate how urban quality is bound up with, and is indeed an outcome of social, economic, physical, historical or other factors.

In post-Thatcherite Britain, the philosophy of an overriding development sector carrying all before it is being replaced (or at least modified) at central government level by a greater willingness to consider the needs and views of the public. The aftermath of the development bullishness of the 1980s, followed by the recession of the 1990s, has given us a good opportunity to reconsider the whole question of urban quality in terms of both failure and success. Most approaches to urban quality literature, however, have tended to be partial and specialised, rather than general and pluralistic. It seems that the time is ripe for a fresh approach.

THE ROLE OF DESIGN GUIDANCE

Many members of the public, either as individuals or collectively, express discontent and unease nowadays with the form in which new buildings or changes to existing buildings take place – apparently not always with the necessary degree of public influence or control. We are already entering here upon sensitive and dangerous ground. *The problem does indeed appear to be one of lack of perception of the contribution and place of urban design in how we control our environment.* There is consequently also a problem of its financial provision and human resourcing, essentially of course via local government and the statutory planning system, which exists not only to 'enable' but to 'ensure'. There is today encouraging evidence that many private sector developers are prepared to contribute meaningfully to enhancement of the urban environment (for example, in the form of 'planning gain' arrangements), though it would be unfortunate if this were the only means open to local authorities to improve the different parts of their towns that are in need of enhancement.

Since, however, so much of what needs to be

(or is) changed in the urban scene is not the isolated prerogative of the private sector or its architects, but has perforce to pass through the statutory system of Town Planning and other parts of the local government machinery, it seems a worthwhile objective to try to bridge that gap. Co-operation and greater understanding on all sides is vital to success and progress. The aim of this book is to try to attain this objective (despite the very wide span and range of the subject) by suggesting insights into dealing with typical situations via examination of context, thus attaining clearer concepts, higher standards and more successful results in the design of buildings and street scenes.

Design Literature[1]

A great many books, some seminal, have already been published on the subject of architectural/urban design and enhancement. In contrast to Prince Charles's well-known book *A Vision of Britain*, two earlier works come to mind (albeit written mainly for architects). These are Gordon Cullen's splendid and comprehensive book *Townscape* (still widely applicable today) and Frederick Gibberd's equally monumental *Town Design*, both of which gave architects and planners much food for thought during the thrusting years of the 1960s, when a major wave of development, comparable in scale to that of the 1980s, all but immersed us.[2]

Over the years, a great number of books have been produced especially for the layperson or amateur enthusiast upon the subject of historical/traditional architecture and its contribution to our towns and cities whilst avoiding modern urban design matters. Others have been published chiefly it seems for practising architects or 'urban designers', and oriented mainly towards the private sector of building and development activity. It appears, however, that there is a special need for publications dealing with the management of urban design and the promotion of environmental quality specifically for public sector planning officers, members of planning committees, members of amenity societies, the general public and (last but not least) planning applicants. All of these persons, either singly or collectively, exert varying degrees of influence and control upon the shaping of our environment, both via the statutory town planning system and the development arena. Our contention is that in promoting a wider understanding of urban design (including promotion of environmental quality), the basic aim should be one of positive urban management rather than negative 'control', i.e. the approach should be one of negotiation where appropriate, and of maximisation of potential, rather than one of frustrating or blocking the natural process of change in the urban scene, where potential for improvement exists but where problems get in the way of realisation.

This book, therefore, in addition to describing for the student the most commonly encountered urban situations and problems (at the same time providing a personal critique of much that is wrong with our towns and cities), is also aimed at the concerns of both the ordinary citizen and professionals, whether as members of the general public, as planning applicants, objectors, local authority members or officers. Its theme of urban quality, and its concomitant theme of the need for design advice and control, should be seen to a large extent as being directly woven into and echoed by the development control process in planning. Also involved in this process are the voluntary sector amenity/preservation societies, as well as developers and their architects, plus (last but not least) the funding institutions, whose actions affect us all. Generally, therefore, the book is intended as a source for trainee planners at all levels, plus architects and all others interested in the quality of our built environment, including non-professionals.

Our study of urban quality is basically a progression from the general to the particular.

It is structured first with an Introduction (Part I) starting with a historical review and leading into an examination of the case for urban quality and the role of design guidance today. Part II then examines selectively the UK town planning framework, via relevant aspects of the development plan and control system; the development process and further measures for guidance of design, development and urban enhancement. Publicity and controversy are looked at, with topical examples (such as the sagas of the National Gallery and Mansion House developments, with their royal dimensions), followed by a look at the related matters of lay and professional education in planning and design.

In Part III we attempt to provide a fresh perspective on urban quality, first via investment in urban quality as the basis for a way forward. We discuss questions of definition/meaning of urban quality. We then explore the fundamental planning and design philosophy of context, followed by an examination of development quality, plus a review of, and commentary on, the essential but devalued role of the public realm. We continue Part III under the heading 'Urban Situations: Casebook' with case references to illustrative studies of urban quality, selected to show the ever-present problems of compatibility of character and scale with existing context in city centres, medium/small towns, historic areas and other situations.

This is followed by a commentary on modern urban design, and on modern architectural design and technology within existing urban contexts. We return to the theme of public controversy with an in-depth look at 'The Great Divide' between old and new. We follow this by considering the use of 'Indicators, Generators and Criteria' as tools for the assessment and achievement of urban quality. The final section of Part III appropriately is entitled 'The Way Ahead in Planning and Development', with specific regard to the very topical aspects of sustainability in urban quality (and its related topic of traffic pressures), and to the need for a revised approach to funding in land acquisition and construction realisation in the built environment.

We have appended to the text where appropriate the main conclusions of our study in the form of generalised recommendations, and at the end of the volume there is a bibliography of reading and reference literature for each of the three main parts of the book. Wherever practicable and constructive, we provide references, while the range of illustrations has been chosen to reflect as far as possible salient themes and relevant aspects of the text.

M.A.B. Parfect
Lewes, East Sussex
G.C. Power
Chew Stoke, Bristol, Avon

PART I:
INTRODUCTION

HISTORICAL REVIEW/SETTING THE SCENE

Society and Environment in Disarray: the Tyranny of the Car

Many architects and town planners studying and practising in the 1960s owe much of their feeling for 'townscape' to Gordon Cullen's copiously illustrated and very practical work by that name, which opened the eyes of a whole generation to the 'art of seeing' the three-dimensional picture of a built environment, and even of appreciating the 'fourth dimension of time' in a two-dimensional plan or sketch of a fragment of town or village. Through his careful analysis of the scene that was almost too familiar, he taught us all to seek (often for the first time) those elusive elements of scale, texture, solid, void and the immense value of space, which are the basic components of any urban experience.

In his photographs and vivid sketches, he was at pains to show the participation of human activity within these elements, and confirmed the urban experience to be the most basic foundation of civilised living. Civilisation created the city, which in turn, through its development and use by its people, became in itself a civilising influence throughout history. The conditions for its success, however, have always required a strong, cohesive social and economic order, contributing positively to quality of urban life.

Looking at Cullen's illustrations today, one is struck by a grand omission from virtually all of his street scenes. There are few, if any, vehicles to be seen. His texts discuss in detail and at length the forms, colours, dispositions and juxtapositions of buildings, and the spaces generated about them, while his suggestions for future re-planning improvements, in a most practical way, consider the alleviation of road traffic from the visual scene. It becomes obvious, however, that the deterioration of the British city scene which we see everywhere today through the almost uncontrolled incursion of the private motor car and the heavy trucker vehicle, was neither felt nor hardly imagined forty years ago. All this seems symptomatic of the overall disarray of our social, economic and political framework, and thus of our environment.

The general decline in Britain's economic life, with towns and cities desperate for new development at almost any price, and the attentions of developers, seeing rebuilding and new construction in our urban environments often as merely a means of generating inflation-proof financial capital return, have not alone created the present damage to our city and town centres. Exponential traffic growth has affected even more both the distribution of land values and the physical condition of our active urban cores. The city itself now seems to be in a state

8 PLANNING FOR URBAN QUALITY

TOWN

rock-bottom subtopia

The two views below, although imaginary, are all too familiar. Wildernesses that in 1956 pass for town and country. How did they get like this? How can they be reclaimed? The TECHNIQUE of reclaiming the wilderness and creating true urbanity and rurality is demonstrated in the following sequence of drawings.

COUNTRY

1 restore the unities

The first principle is respect for the category. Ask the question: What am I dealing with? Wild, country, arcadia, town or metropolis? Having decided, stick to the category and augment its proper character. Town: remove suburban rockery roundabout and hoarding decoration; respect architectural integrity in shopfronts. Country: informal road and footpath are adequate: replace suburban rockery and planting by mixed hedge, remove urban advertising.

Figure 1 Extract from Counter-Attack: urban design versus 'subtopia'.
In the post-war period of the late 1950s/early 1960s, degradation of the visual environment was already evident, and was addressed by Gordon Cullen and Ian Nairn in *Townscape*, *Outrage*, and *Counter-Attack*. We start with an extract from one of these seminal works.

INTRODUCTION 9

TOWN

2 remove the clutter

COUNTRY

Second, make sure that every single item that is seen by the eye is essential. Town and country are not dumping grounds for reiterated instructions, most of which are unnecessary. Above all avoid the use of verticals. Town: keep light fittings slender and site unobtrusive, keep road signs to minimum. Country: where possible bury wires; use road surface and walls for road signs and bus stops.

3 camouflage and bring together

Third, don't waste space. Town: fill in vacant corners with buildings, not gardens: bring back enclosure as opposed to impersonal emptiness. Reorganise traffic so that the dead space in the system is returned to pedestrian use. Country: gather up the scattered buildings, thus extending the area of visible countryside, and at the same time creating a recognisable place, instead of a piece of sprawl. Finally, camouflage the interruptions that can't be moved.

The sequence shows the brutalising effects of a 'subtopian' approach to street furniture and traffic signing in both urban and rural scenes, and how sensitive and intelligent design measures can remedy the situation.
Source: Ian Nairn, *Counter-Attack*, Architectural Press, 1957

of rapid disintegration because of this impact of motor traffic. Until this most damaging factor in the achievement of satisfactory environmental conditions is addressed, much of the guidance one might give in town planning terms towards the achievement of an urban quality in many of our town and city centres must remain as an expression of hope rather than a recipe for rapid improvement.

As this is written, road widening and many other traffic management schemes, with their plethora of signs and road surface symbols, are being continually 'shoe-horned' into the old fabric of our town centres. Huge superstores are elevated in eye-raping colour combinations of plastic-covered metal cladding (to attract the shopper travelling by car, no longer a pedestrian). Adolescent tree-planting schemes attempt to improve or conceal commercial car parks while outer suburb street parking mushrooms in newer and older housing developments alike, as the two-car family changes inexorably to the multi-car family. The recent major growth of traffic-calming schemes in the UK, whilst being a brave but costly attempt to mitigate the harmful effects of traffic upon the urban environment, is indicative of the problem. At trunk road and motorway junctions, out-of-scale pitched-roof office blocks masquerading as glazed barns, church halls or ungainly and ill-sited country mansions float in water features, and in lakes of parked cars. We all begin to wonder if townscape itself has not indeed been bypassed by the impact of 'motor city' on the move!

In the old city centres, the pavements crack under the load of vehicles where once pedestrians walked in some safety, and the streets become deserted of people who travel in 'transports of delight' by favourite commuter routes out to the suburbanised villages and the ruralised high-density tract estates of design-guided housing built in materials brought easily by road from any part of Western Europe. Midlands red brick in Devon, Danish windows in Wales, tropical hardwood everywhere only make sense in a Britain driven by the motor vehicle, through run-down local industries anxious to diversify.[3] Television's edited lens will show us other environments where sleepy medieval town squares, elegant seaside terraces, quiet fishing ports and car-free open spaces beckon us to our expensive holidays abroad. It's good to get away from it all, at least once a year.

It is a folly, which is in imminent reach of causing irreversible damage to virtually all our urban and many of our rural surroundings, when the motor vehicle, adaptable to change in its own short life-cycle, should be allowed and even tacitly encouraged to destroy our long-term investment in the created environment. We are coming to terms with the damage similarly inflicted on many parts of Britain by the excesses of urbanisation. Belatedly we have learned that the uncontrolled pursuit of short-term profit and convenience leaves a long-term legacy of ill-health, substandard environment and economic inefficiency in train. The impact of the car and truck is already more universal and harmful than the pollution of the 19th-century 'Satanic Mills' and is a great deal more insidious in its effects upon the national lifestyle and mores. The anonymity of 'sealed-box' individual transport at almost every opportunity of travel, and the consequent reduction of personal contact, has contributed to a diminishing social consciousness and a lack of community awareness, unknown even in the 'tunnel-back' housing of the industrial work-camps of the last century. The very observation of our day-to-day surroundings has been devalued by the superficial view obtained at 50 kph and faster. We do not walk about our environments in Britain as we once did, and this is having a profound effect upon our perception of unsightliness in our surroundings. The growth of street litter, ill-maintained building and street fabric, garish or simply neglected sign-writing in the general scene, and derelict urban

INTRODUCTION 11

Figure 2 Dublin street: visual qualities.
A 'snapshot in time' of Dublin in the 1990s: all is in balance. Man, cars, plus architecture from many European styles and epochs, successfully co-exist. Treescape is dovetailed into an urban environment of dignity and calm which has been produced by human energy informed by sensitivity. (Contrast with the 'subtopia' of Figure 1.)

There is sufficient intimation of Arcadia here with temple, portico and trees to incorporate this into the city without making the mistake of trying to realise it to the exclusion of all else that a city must be. *The effectiveness of the fragment* is thus illustrated. Well-designed architecture, from different periods, but sympathetic in scale, linked and punctuated by recesses in the 'street wall' are not dominated by the trees. The gently curving street also lends grace and anticipation to the vista, and the generous width of the street absorbs kerbside parking as a decorative accent to the general scene.
Source: Clare Power

sites used as rubbish deposits are not only indications of economic malaise but also a symptom of a society that is alienated from its surroundings and, frighteningly, that is not even seeing the effects of general neglect any longer.[4]

Town planning today in Britain has a much more demanding task on its hands in the restoration of urban quality than it had in the aftermath of war in 1947, when the pioneering Town and Country Planning Act came into effect. Our vision of Britain at the end of the 20th century is all too heavily clouded by vested interests and entrenched practices and policies for any rapid or widespread improvement to urban quality, unless a fresh approach is made – even at this late stage – to its essential importance in shaping our lives.

Historical Perspective: Background to Urban Quality in Britain

Until the economic expansion of Britain following the era of Elizabeth I, and the profound social change in the wake of the English Civil War, which provided much of the impetus for the phenomenal trading and industrial growth of the next 200 years, the English had no urban tradition in the European sense. Wealth, such as it was, comfortably spread across rural centres amongst some 4 million people after the Black Death of 1348, was concentrated in land and the widespread villages, still to be seen as archetypal 'Olde England'.

Up to the 19th century, Britain was not in the forefront of European efforts towards the realisation of designed urban development. Even then the grand concepts of Haussmann in Paris and Cerda in Barcelona did not appear in our cities. We preferred to link our villages to produce our conurbations. Grand planning concepts usually involve too many small freeholders of land in selling their hard-won titles. Even today the assembly of land areas to accommodate large planning concepts in high-value urban areas remains a stubborn barrier unless sizeable tracts of publicly owned land such as green open spaces can conveniently be purchased and developed for private sector purposes, as is the present trend.

The Roman idea of the city, well rooted across the lands bordering the Mediterranean basin, came relatively late to the British Isles. Although London, York, Chester, Bath and vanished Silchester provided worthy small examples of the Classical idea behind the planning of practical, high-density, self-sufficient environments for well-organised urban societies, the Western European ideas of *rus* and *urbs* (as distinct and complementary experiences in living) did not find a ready reception in the British way of life.

It is true that the Roman occupation brought something of the ethic of Western European urbanisation to Britain, as its legal systems and social customs were imposed upon a scattered population of probably no more than 400,000 country dwellers at the start of the period. It is interesting that whilst most Roman centres were abandoned to decay during the long Dark Ages period, and in fact were neglected soon after the Romans left Britain, those few cities that were reoccupied (such as York, Chester and Bath) have about them, even today, a certain feeling in their old street patterns and central area scales, an atmosphere of an urbanity, which has taken root across Europe in a much more positive way than the incursions of Britain's later Germanic invaders allowed, during the imposition of their own lifestyles here.[5]

The advanced technology of the Roman way of living, involving theatres, gymnasiums, law courts, baths, central heating, piped water and drainage together with paved highways, forts and small towns, all systematically planned, returned but slowly. Indeed, it was not until the end of the 18th century when the need for higher standards became more acute, that the comfort and convenience of urban living once again began to achieve the former Roman level.

The historic development of towns and cities,[6] at road and river crossings, fortresses, pilgrimage centres and ports, was always based on the assumption that such centres would encourage trade and wealth for kings and religious houses, but the Middle Ages saw many failures in these attempts to found such urban environments arbitrarily on 'green field' sites. The new town of Salisbury was an exception to this common hit-and-miss activity, being well laid out, on a canal drainage system, and economically successful to the present day. Edward I was not so fortunate with his grant of a charter in 1286 to Nova Villa across the harbour from present-day Poole. Although this provided a most promising new trading site for the growing links with France at that time, nothing came of the project, and it is a salutary lesson

Figure 3 Medieval towns: Lincoln.
Medieval European towns were often centred on a church, castle or cathedral, as well as on the marketplace, as seen in the English cathedral city of Lincoln. They have retained an intensely human character and scale, well fitted to the needs of the pedestrian today. The Continental town displays, through its evolution, a slightly more formalised character, as against the English informality.
Source: Michael Parfect

for all who would plan new towns, that natural growth often depends on factors that are uncertain in their prediction, especially when populations are stable, or increasing only slowly, as is the case in Britain today.

Castle, cathedral and monastery formed the early medieval urban experience for the emergent English. Meanwhile the economic expansion of overseas trading (in spite of the population decimation caused by the Black Death across Europe and England) made the bourgeois life something to be admired for its comforts, won from dealing in wool, wine, silks, spices and shipping which opened up the trading routes of the world. Towns and cities from Lombardy to the Low Countries, and throughout Britain served the land and sea routes of great economic growth. City walls and protective bulwarks provided the lanes and later boulevards of the 16th and 17th centuries. Wheeled traffic replaced the pack horses and sleds, and the lanes widened to streets and roads. The city as a place to be, to live, to learn and to grow in, became preferable for a rapidly expanding population. Thus, by the time that the Woods (father and son) of York were indulging the wealthy dilettanti through the introduction at Bath of a concept of an imagined Ancient Rome, the architectural backcloth it provided in terms of the social, cultural, classically planned layout of the 'pleasure city' was already beginning to address the cascading problems of high-density, terraced housing and work places for a country people deserting the isolation of rural existence for the dirt and financial deliverance of the 'industrial city'.

Much of the traditional urban quality in Britain relies upon the experience of a population which grew from some 5 million at the end of the 17th century to 48 million in 1951. The 'ideal towns' of the Italian 'Quattrocento' translated briefly by Wren, Hooke and other 'New Men' of the late 17th and early 18th centuries in Britain, were largely set aside (regrettably but inevitably) in the race for industrial growth and its consequent wealth. The avenues planned by the architect-planners of post-revolutionary England, such as first the Woods of York and Bath (building elegant squares and crescents on gentlemen's images of Roman antiquity picked up in the enlightenments of their Grand Tours), soon became the tight gridiron layouts of new towns such as Preston and Manchester where country people, displaced at last from the growing starvation of an overcrowded rural existence, went to work the machinery of the Industrial Revolution.

Compared with what had been experienced previously in England, these newly built towns, planned in logical relationships to work places and new roads, were desirable places in which large families could earn financial rewards, and from which they could travel to other places, as the railways formed new market towns and mercantile centres. The fear of cholera and plague was conquered by Victorian engineering expertise, and the industrial urbanism of a burgeoning Britain provided the first conscious experience of the dangers and delights of city life for the bulk of the population. Seaside resorts provided further high-density living experiences out of the industrial 'smokelands' but the concept of a 'New Jerusalem' based on prosperous, democratically run, large urban environments, set in freely accessible countryside, perished in the economic decline which signalled the Great War.

William Blake doubted that the cotton mills elevated the spirit as much as they increased the wealth of the nation, but both he and artists like Turner found great romance in the reddened sunsets and the smoking chimneys set against the blackened skies of the age of steam. New hospitals, railway stations, workhouses and orphanages, town halls and the public libraries, metalled roads, and sanitary recreation parks became the new adornments of a popular urban environment. The slow-moving world of gradual change and the picturesque was left behind in favour of the functional urban machine, designed not only for

Figure 4 Baroque city: Rome.
The development of Baroque city planning in Italy took an altogether more overt and powerful form than in medieval England, France or Germany. Human activity and presence were strongly reflected via formal civic spaces, some curving or tapering in plan, utilising grand axial streets as approaches or linkages between spaces. But the underlying intent was for control as well as for visual effect.
Source: Christine Parfect

16 PLANNING FOR URBAN QUALITY

INTRODUCTION 17

hard work and hard living, but also for hard leisure and entertainment, bought with the ready money that needed to be earned every day.

Health suffered; happiness suffered. By the beginning of the 20th century the work was running out, the environment of many of the new cities was blackened and ruined. The rural background was decayed and impoverished. The old way of life was gone for ever, and two world wars meant that a new way had to be found for people to live in some sort of health, wealth and happiness, with time to enjoy the quality of their environment.

The idealism following the aftermath of inter-war slump, and a victory for liberal democracy in 1945, produced the desire to rediscover a secular 'Jerusalem' in a planned economy, planned education, planned working and the Town and Country Planning Act of 1947. Britain, it was felt, would regain its former prosperity; the population would grow and become healthier and wiser; and the new Utopia of the 20th century would be realised upon the foundations and the remains of the 17th, 18th and 19th centuries. Nuclear electrical energy, automated production at all levels, and the freedom of personal mobility (in the form of the private motor car), would ensure lasting happiness, health and prosperity, in terms of new buildings for all, with only the very best of the old conserved. *For many reasons we do not appear to have achieved this happy result at the end of the 20th century.*

The programme for regeneration in the physical surroundings of our lives, our working

Figure 6 20th-century motorcity: London.
This illustrates how general needs of the city as a whole can compromise the integrity of a locality. Elephant and Castle, geographically at the centre of London, is shown first in Goldfinger's vision of a contemporary public space. Its reality today as a traffic roundabout fragments its coherence for the pedestrian and alienates the lives of the surrounding buildings from the central core.
Source: Mark Power

Figure 5 (opposite) 19th-century city: Barcelona.
The 19th century saw city planning developed further to meet the needs and pressures of a burgeoning but overcrowded population. In Cerda's plan for the expansion of Barcelona, the medieval core is very largely left aside and preserved, while the vast bulk of the new city structure follows Cerda's grid layout. Note, however, the design refinements of the grid – the space formed at each road intersection by block corner splays, the elliptical roundabout and its curved block responds, and the generous tree planting. Only the progressive block infilling by developers has compromised Cerda's farsighted and sensitive urban layout plan.
Source: Aerial photograph of Barcelona. Reproduced courtesy of the Institut Cartogràfic de Catalunya

Figure 7 The city experienced from the vehicle.
Two contrasting experiences of the city through public and private transport. On a bus the city is experienced as a passive panorama, and freedom of movement is sacrificed for the shared journey. From a car the city is seen ambiguously as, on the one hand, a labyrinthine series of obstacles to be negotiated en route to your destination, yet on the other, subject to your control, held within a window above a panel of instruments. Common to both is the state of insulation and detachment from the life of the streets.
Source: Mark Power

conditions and recreations, appears at the present time to be a daunting one, but it is not unique in town planning history. The late 19th century saw similar problems across Europe as outdated, often medieval major cities strove to modernise and their own social and industrial revolutions began to follow the model perceived to be so successful in Britain.

The long-established urban tradition of European cities faced a much more critical problem of renewal, for example in Berlin, where the influx of new working populations had not been housed in the comparative comfort of northern Britain's new urbanised 'work camp' towns. Instead they had merely been obliged to set up homes in shanties outside the old city perimeter, or, as in the medieval remains of old Paris, to work out an ever more crowded existence in the decayed remnants of insanitary warrens of once respectable old town houses. Individual landlords saw no future in piecemeal renewal and the Paris mob, remembering its power to survive in the narrow streets, provided a slow fuse of ugly social unrest, as recorded in the novels of Victor Hugo and Emile Zola.

Against this background of patent urban failure the enterprise of Baron Eugène Georges Haussmann achieved unparalleled success in the Paris of the mid-19th century. By ruthlessly arranging for the buying up and bold replanning of the whole of the old city, he was able to provide the boulevards, avenues and a beautifully elevated, instantly landscaped and socially safe, modern city style, emulated throughout Europe and much of the Latin world afterwards. Haussmann proved that redevelopment was not only capable of financing itself, but also actually able to increase value and amenity at the same time, and in varying degrees his formula held good wherever it was applied, for instance by Cerda in Barcelona, where a major city expansion was successfully and sensitively added on to the medieval core whilst still preserving its integrity.

There is little doubt that creative development of this kind can be achieved in urban renewal today, where will, speed of acquisition, and short construction time can all be assisted by political intervention. It is axiomatic in our modern democratic society that honesty of development purpose, realisable to a strict, forward timetable, should lie at the root of all economically successful building development. It has not been to the advantage of British society that the realisation of very necessary urban redevelopment projects, in a chronically high-interest cost era, has quite often been thwarted by lack of confidence and the political prevarication which it breeds.

For the British then, the urban experience has been relatively short, and in comparison with much of Europe, of an almost artificial quality. *The great engine of wealth which generated many of our principal cities in the 19th century has run down, as the need to renew has arrived.* Our images of the new 'Ideal City' are either academic or based upon criteria that are no longer relevant. Our ideas are becoming gradually world-based, but our relative income in world trade terms is tending to be again almost that of an off-shore, semi-rural community, forced, as in Elizabeth I's time, to live by the wits and energy of its people. Nor can we afford to turn our back on the European Community of the future. We must regard the future realisation of economic urban quality against a background of spent achievements which need to be recast in the experience of modern conditions and desires, to provide a workable design for a fulfilling lifestyle, rather than as the rearrangement of most of the lumber from a rich but varied past into a somewhat meaningless museum for an aimless population.

THE CASE FOR URBAN QUALITY

The Need for Promotion via Local Government

There are a number of major issues of political sensitivity and concern which, by any standards, ought to occupy an important place in both the national conscience and the governmental system of Britain. They are respectively those of economic prosperity and fair distribution of wealth; a comprehensive and just legal system; provision for education, health, and general social well-being; and lastly a high quality of environment.

Whilst the other issues have always been to different degrees popular preoccupations, the same could hardly be said of the last issue, except in very recent times. Too often it has been dislodged out of the frame of governmental consciousness by the adversarial way in which this country's affairs unfortunately are run.

Now, the after-effects of unusually acute pressures of development experienced until recently in many parts of the country (caused by the free-market economics of the 1980s) and the severe restrictions on the role of local government in controlling these activities, have stemmed from the tendency (up to the more recent property market depression) for the environment to be either overheated or neglected, depending on the particular location. It has to be emphasised that the British have never been quite as environmentally conscious in the management of their towns and cities as some of their more progressive West European counterparts. Our somewhat blinkered national pre-occupation with other issues (particularly financial and legal) does not on the whole serve to focus attention upon the need to promote enhanced urban quality within the overall range of our towns and cities, where most people live. It is not, generally speaking, for lack of potentially available professional expertise in this country; it is most often for lack of the means or perception (at either national or local level) or simply the political will to harness that expertise in an informed manner to deal with the problems on the ground. In this, *the role that needs to be played by local government in its administration of the planning system in this country is vital – but unhappily is too often underestimated.* Yet the knowledge and human resources are potentially available in this country; more particularly, however, from within the private sector. Local government, as traditional guardian of the local environment, needs the political will and the financial and human resources to play its full and proper part in ensuring appropriate urban quality and design, beyond a merely restrictive system of control (a concept with which it is all too often stigmatised). The obsessive grip exerted on the public sector by central government through the 1980s, however, gave local councils little scope or incentive to spearhead this process.

Urban design advice – essential to good urban environmental quality – could well be 'externalised', i.e. carried out by the private sector, in this way. On the other hand, it would seem both illogical and impractical to privatise development control (the 'flagship' of Town Planning at local level), due to overriding political questions of local accountability. Certainly, there is a clear need for urban design advice to be available either externally or internally to local councils as a specialist adjunct to their planning control systems.[7]

People rightly demand not only protection of their environment (both urban and rural) but also its enhancement wherever practicable. The urban scene needs protection against a number of ills, including the frequently interlinked factors of overdevelopment and traffic congestion. Also interlinked, in visual terms, are the twin qualities of overall townscape and individual buildings. New or changed buildings can (and often do) have a disproportionate effect upon overall town quality. It is, therefore, directly incumbent upon developers and their professional agents (principally architects, but also surveyors and commercial advisers) to put forward building schemes that are environmentally sympathetic in every possible way. The urban fabric of the towns and cities of our overcrowded island is too sensitive for anything less to be acceptable. At the same time, local authorities and local communities have at least an equally pressing concern to ensure that the changes occurring to their environment are well judged in principle and of the best possible quality in detail. An intemperate or egotistical scheme by an insensitive designer can have just as disruptive an effect upon the urban scene as an overblown scheme of excessive content put forward by an overambitious developer. Major prestige buildings (which nowadays seem to be the preoccupation of a central coterie of British architects) are not, generally speaking, the issue. The main issue now facing us is that of local communities assimilating urban change in acceptable forms within their town centres and residential areas. The evidence of that need is all around us. Yet blinkered thought and piecemeal parochialism, with a strong undertow of adversarial attitudes, prevail within the decision-making process.

The Prince of Wales, in his book *Vision of Britain* publicly voiced his concern, in presenting his own personal views of some of the main defects. Unfortunately, whilst stirring up a hornet's nest, he has posed more problems than solutions. A balance has to be struck, meanwhile, between maintaining traditional values and allowing bona fide change and improvement to take place. Local authorities, the private sector and local people have a rightful and central role to play in this process, relative to urban design, quality and control. They must be assisted, not hindered, in playing that essential role, by central government in particular.

In this context, it is temptingly easy to pontificate as to how much the public sector and the planning system can or should do to promote urban quality and good design. The very nature of planning and the way in which the British development market operates, however, make this task a difficult and indeterminate one. The central function of planning within its forward-looking role is to regulate change, in terms of maximising its potential for the public good, and at the same time to minimise any attendant harm. The central capability must be that of controlling land use and of shaping the physical environment.

A major underlying difficulty, however, which besets the local planning authority at almost every turn, is that of multiple private ownership and value expectancy of the land in question. This aspect, amongst others, inhibits planning authorities in Britain from replacing restrictive control with positive and constructive creativity. *A system based mainly on what planning prevents is not likely to be as successful in environmental terms as one that actively promotes high standards of urban design. Yet it is the more negative of these two roles which is still predominant in the minds of developers and the public.*

How far can one get away from this highly unfortunate concept? In the minds of many people, planning owes its existence basically to the presumption that land may normally be developed – without such intrinsic force, there is no planning to be done in the first place. Although the local planning authority now has very little scope for developing its own land, it has the residual if limited role of allowing private building to take place hopefully within the framework of planned urban concepts and sometimes with financial inducements. It also has a role promoting physical enhancements hand in hand with developers or with local business communities and residents.

There is thus a basic dilemma, whose satisfactory solution must to some extent rely on the developer's wish to realise a high-value scheme, plus the landowner's anxiety to avoid escalation of development costs through delays in obtaining the necessary full planning consent – and on their joint willingness accordingly to go (at least some way) along the road of design negotiation and improvement. Although this may in many cases amount only to a departure from the substandard humdrum, or ordinary, in favour of something more interesting and acceptable (rather than the attainment of great heights of architectural design), *it can in fact represent a useful move in the right direction and should not be discounted. Not every development can or need be a tour de force of either modern or traditional design.* It is, however, in respect of the 'collective control', upkeep, improvement and embellishment (and frequent repair) of the urban 'tapestry' that local planning authorities still have just as important a role to play as in the more generalised abstractions of policy-making for the future – especially when the future can be as volatile as we know from experience it can be. It has to be said that planning policy decisions have sometimes been based on incorrect assumptions, have at times been lacking in foresight or flexibility of view, have been negated by political U-turns, or have simply been overtaken by events.

As Michael Middleton says in his excellent book *Man Made the Town*:

> All planning is a matter of trade-offs. There are many routes to many solutions, but there is never any perfect solution. . . . *In the last analysis, the machinery of planning is turned by attitudes, vision, and political will.* If our environmental programmes are too often tentative, short-term, and lacking in consistency, it is because society itself remains uncertain as to its real objectives, and its leaders too often fail to foresee the secondary effects of their decisions.
> (Middleton 1987: 87, our emphasis)

It is the aim of this joint study to examine the problems underlying and inhibiting the creation of greater urban quality in our environment, in which context our lives are shaped and lived.

The Planning Situation Today

The past fifteen years have seen an enormous upsurge of development and urban change in this country, on a scale that has led unsurprisingly to growing public concern at the effect of the amount, form and location of all this physical change within the environment. In particular, people are understandably worried about the quality of design of new buildings in the urban scene, and they look to local planning authorities increasingly to ensure that the necessary expertise is available in all cases to deal in an enlightened and effective way with development proposals and planning applications. At the same time, there is also a general expectation by developers and applicants that new development should furnish a healthy economic return for the initial financial outlay or investment. This places a major responsibility upon local authority planning officers and members of planning committees to ensure that a high standard of understanding and judgement is applied to that process.

Design Guidance

In this context, there is virtually nothing of greater potential for public controversy than the design aspect – yet there is scarcely any aspect so little understood by the ordinary person and at the same time so marginally provided for within the ever-diminishing powers and financial and professional resources of most local authorities. The same can be said in respect of enhancement of areas of the physical environment, where the opportunity for this arises.

The fault, generally speaking, is not that of local government. District and county councils in Britain have for some years now been making determined strides to redress the balance in terms of providing their electorate with adequate professional resources and advice on design matters. Firm and sensitive design guidance has been provided wherever possible to applicants and council committees alike, and private consultants also employed where town centre redevelopment or enhancement studies are needed by local authorities to improve the areas for which they are responsible. *But the system is often under-resourced and overstretched.*[8]

Central government, though apparently more aware now of the major importance of the environment including urban design, exerted throughout the 1980s a stringent squeeze upon local authority finances, and thus upon the professional resources affordable for public sector town planning functions generally. There has been a real danger of the baby being thrown out with the bathwater. In a situation where the statutory aspect of service to the public (i.e. those elements required by legislation) must always come first, and where money is consistently tight, functions which might at first glance seem to be peripheral or non-essential, such as urban design and environmental improvement are not going to be able to claim a large part of most local councils' budgets unless the funding basis for local authorities can be widened. *The responsibility of government cannot be avoided in all this, and it is high time that this situation is fully recognised at governmental level.*[9]

Another stumbling block is the vexed question of 'design control' by local authorities. Some of those who operate in the private sector (developers, architects in particular, and private applicants generally) from time to time question the right or competence of local authority planners and planning committees to exert a degree of design control. There are many instances where this is exerted with good reason. This generic term can imply a whole range of things, from the active participation of planning officers in the draft proposals stage (prior to the applicant's formal submission of a scheme for planning consent) to outright rejection of the formal proposals at committee stage.[10] Up until the mid-1990s the Department of the Environment rarely supported on appeal a local authority's rejection of a scheme purely on design grounds (indeed, they would often look for any possible means to approve it, sometimes setting aside local plan policies to do so). Unless the proposals were manifestly unsuitable and the environmental setting highly sensitive, the presumption on appeal was normally towards approval. Councils were accordingly reluctant to reject a scheme purely on design terms (especially with the shadow of adverse cost awards for insubstantial reasons for refusal looming ever larger) unless they were prepared to 'chance their arm'. For that to succeed they had to have sufficiently strong and informed advice from their officers, and be clearly enough convinced in their own minds to refuse the design proposals or to call for design negotiations and amendments. Planning officers, for their part, were often reluctant either to demand changes to schemes, or to recommend their committees to reject proposals, unless they felt very sure of their ground professionally.

The whole thrust from the DoE has now changed; PPG1 (General Policy and Principles) states that 'Development Plans and guidance for particular areas and sites should provide applicants with clear indications of local authorities' design expectations.' Reinforcement of this view was provided via the revision of PPG1 in 1996, and this will help planning authorities in dealing with questions of design control.

A particular hazard, especially subtle, which many experienced development control officers still tend to steer clear of, has been that of acknowledging the design merits of a developer's proposals where from the outset clear conflicts with other established planning policies (such as permissible floorspace) exist. They know that if they give such aid and comfort to the developer, yet the proposals are not amended or reduced to comply with policies, and are subsequently refused and go to public inquiry on appeal, the officers are quite likely to have their earlier encouragement of good design standards used against them and large wedges driven in, at the inquiry, in the form of accusations of 'inconsistency' by the skilled legal advocate (often a QC in cases of major proposals) acting for the appellants. This is of course a sorry state of affairs, because it directly inhibits constructive design discussion in the first place, which in turn benefits nobody. But *constructive compromise - a minefield for the authority — is sometimes essential if best results are to be obtained in the design form of the development.*

Planning officers, particularly the more senior staff, bear a centrally important responsibility in dealing with applicants' design proposals, since it is they who must advise their planning committees on the merits or demerits of each case in the end. The more fortunate or enlightened planning authorities will already employ their own urban design and conservation staff (where they are available and can be afforded) to make the necessary design advice input to the development control process. However, these specialist staff, especially architect/planners, are in relatively short supply in local government, which must compete salary-wise with the private sector for them.

The validity of design control as a concept in itself has more than once been brought into question by members of the architectural profession in particular, from the private sector side of the fence. The argument in favour of design control from the viewpoint of benefiting and safeguarding the public's wide interests has been rehearsed above. The argument against this concept, as put forward by some private sector architects, is basically an opposition towards 'amateurs and laymen meddling in the rightful province of design professionals'. Whatever sympathy one might have with this view, it is counterbalanced not only by the need to safeguard the public's interests (in the democratic context of local government) but also by the significant fact that only a small minority of the total number of planning applications (about 13% in 1990) is submitted annually by architects. Architectural designers, surveyors, builders, etc. plus private applicants (also unqualified in architecture) make up the remaining 87%, though it has to be said that 'architectural designers' (or 'plansmiths' as they are often unkindly known!) are quite often able to submit adequate and appropriate design schemes, where the nature of the proposals is within their scope – more usually minor domestic schemes. For schemes of a greater range or scale, or for locally sensitive areas, developers will normally employ architects. Developers, however, are primarily interested in the maximum quantity of commercial return from their schemes: this normally means getting as much as they possibly can on a site as quickly as possible, however well disposed they may be towards good design solutions. Sites with 'development value' are expensive!

None the less, intelligent developers are beginning to appreciate that 'good design pays', and their architects' task is thus to produce a financially advantageous and functionally sound yet aesthetically pleasing scheme, well related to its location. A skilled architect will succeed in this task, but hopefully will at the same time be willing to discuss the design and appearance of his or her scheme with the local authority's planning officers, and make changes where appropriate. In this process, he or she may or may not be helped by the developer or client. The worst type of developer will insist on an unremittingly commercial scheme, not only with maximum floorspace, but also of an insensitive design, e.g. dull, heavy, too plain, or out of either scale or character with the site's location. In this circumstance, the architect may often need a degree of protection from the client by the local authority planners (a situation which occasionally is tacitly expressed to the planners by the architect!). The reverse can also be true, where an unduly idealistic architect/designer is involved.

So not all is 'plain sailing'. In any situation that results in a design scheme of doubtful or of 'overblown' character or content, it would seem that the local authority has the right and duty itself to intervene and try to assist constructively – whatever the status of the designer or the underlying factors. Few architects would surely quarrel with that, as long as the advice they were getting from the planners was informed professional advice and not unfounded preferences or opinions, e.g. about 'style' (which can in many cases be a particularly misleading approach).

It is recognised that councils should, wherever appropriate or possible, produce planning briefs *for the guidance of intending applicants as to the general principles of development in the particular planning context of the site, and how the proposals should relate to them, in broad design terms.* Those principles would include such aspects as appropriate height and mass, and should not extend to detailed facade design or architectural appearance. But often it is the details (such as fenestration, proportion or articulation) which contribute directly to the successful outcome of each scheme. However, the current recognition that town planners have a positive role in terms of providing a planning framework to prospective schemes (in the form of development briefs, planning statements and the like) is at least a step forward from the earlier negative and restrictive concept of what planning/development control was all about. The logical conclusion is surely that there should be urban design strategies linked directly to land-use aspects.

On a wider scale, the whole nature and process of urban change and development have become increasingly democratised (though some would say 'bureaucratised'), including the related aspects of design quality and control. This surely should be seen as a step towards a restoration of civic pride, which is an integral aspect of urban self-government. The roles of urban design, enhancement, conservation and control generally thus appear now to be established elements in the whole field of public affairs and opinion.

PART I – SUMMARY/KEY ASPECTS

HISTORICAL REVIEW/SETTING THE SCENE

The theme of this first section is that of 'Society and Environment in Disarray', in which we depict and trace the visual, physical and social decline of Britain, which is currently to be seen in a number of ways. Our towns and cities suffer increasingly from inappropriate development, set in a tidal wave of motor traffic unprecedented 40 years ago ('the Tyranny of the Car'). Society will become ever more alienated from its surroundings unless planning can regain not only the ways and means to control the twin demands of development and traffic, but also the constructive attitude and vision needed to create relevant and attractive living and working conditions in our urban areas.

THE ROLE OF DESIGN GUIDANCE IN PURSUIT OF URBAN QUALITY

- In most matters of urban quality, the role that needs to be played by local government via the town planning system in this country is vital – but unhappily too often underestimated on all sides.

- Private sector developers and their architects/agents have a duty to provide building schemes that are environmentally sympathetic in every way – local authorities and the public alike have every right to expect and require this.

- Central government has a key role to play in enabling this to happen, but not simply via the indiscriminate operation of market forces.

- A system based chiefly on what planning *prevents* is not likely to be as successful as one that actively promotes high standards of environmental quality. Protection of land values must not preclude the latter course, and the planning system should be operated accordingly. Pre-application negotiations on design quality and improvement potential are one means towards this end. Evidence of use of these should form part of planning committee reports, where applicable, as a material consideration towards the grant of planning consent, and in the determination of planning appeals.

- Middleton in *Man Made the Town* (1987: 87) states that: 'In the last analysis, the machinery of planning is turned by attitudes, vision and political will.' There is a problem of lack of perception of the role of urban design in how we shape our environment. The public have, however, a healthy concern about design standards and the impact of change on the environment.

- There is an ongoing vexed question of design control by local authorities. Its existence is symptomatic of the problem – on both sides.

A process of constructive compromise is essential if best results are to be obtained in the design of development. Unfortunately, design skills are at a premium in local authorities today, but they are still necessary.

- Local councils should, where relevant, produce outline design guidance via planning briefs for particular sites. Planning/design briefs must comply with relevant approved (or draft) local plans. The practice of compiling these in liaison with a prospective developer should, however, be either discouraged or at least treated with caution.

- There should also be urban design strategies linked directly to land use and development, via local plans, etc., produced with the benefit of independent private sector advice.

- Detailed design considerations will inevitably arise, and will be of legitimate concern for public sector planners (though not in an overly prescriptive way as this unduly restricts private sector architects in exercising design skills). The nature of the urban context will be the determining factor in the end, rather than a futile debate based on preconceived notions as to 'style' (a misleading concept).

- The use of local materials wherever possible will go a long way towards ensuring good (sustainable) architectural standards. It is the changing technology of new buildings which often worries the public most, and which should therefore be justified. 'Hi-tech' buildings/architecture in particular must relate satisfactorily with the urban context, especially in an integrative situation. This must be a legitimate consideration in the granting of planning consent.

PART II

TOWN PLANNING FRAMEWORK

CONTEXT AND MACHINERY FOR DESIGN ADVICE

Resources and Scope

There is little doubt of the need for urban design and environmental advice to be available to local authorities as an integral part of the 'mainstream' performance of their statutory planning duties. This need is constantly present in even the most routine aspects of development control, and is not something that should be confined simply to dealing with the more major building proposals, or to conservation areas, or to historic/listed buildings – though clearly the more prominent the site, or the more sensitive the existing building or context, the greater is the need for the local authority to have skilled urban design advice.

This advice can, of course, be either internal, in the form of trained/knowledgeable staff, or external, in the form of specialist architectural/urban design and planning consultants from the private sector. When dealing with prominent historic buildings or ancient monuments, for example, the services of specialist private consultants will normally be necessary. Similarly, detailed planning and design exercises on sensitive/complex urban regeneration or redevelopment areas will also frequently call for the input of skilled consultants.

However, the vast majority of design problems usually occur in respect of 'bread and butter' or day-to-day development control matters, i.e. planning proposals for residential 'infill' schemes or extensions, or routine town centre commercial renewal or redevelopment. 'Hyper prestige' schemes are very much in the minority and are unrepresentational as such. It is in respect of the multitudes of routine or semi-routine cases, as well as of the prestige schemes on prominent sites, that the local authority will need to have its own in-house urban design expertise readily available.

Here we encounter first the resources problem. Whilst money can normally be found by local authorities for the occasional hiring of private consultants for particularly important exercises or projects, not all councils will find it easy or readily justifiable to carry their own urban design/conservation staff as an extra charge upon their ever-tightening staffing establishments. Nevertheless, such is perceived to be the importance nowadays of maintaining and improving the quality of the urban fabric in the eyes of the general public that a strong case can normally be made by local authorities to employ wherever possible some specialist design staff (architects, architect/planners, project engineers and landscape advisory staff, for example) *even at the cost of prioritising this type of expenditure over and above other more discretionary items.*

This leads us directly again to the vexed question of 'design control' and how far local authorities should legitimately go (if at all) in the exercise of this particular function. In the introduction preceding this section, we set out what we consider to be valid justification of local authorities exercising a necessary degree of design control, provided that this is approached in a constructive rather than negative fashion, through joint negotiation, leading to a 'design consensus' wherever possible. In recent years, there has been a series of ministerial pronouncements, via Planning Policy Guidance Notes, as to the role of local authorities in this respect. We shall consider the more relevant of these later in our present study, but Annex B of PPG1 may be identified as the start of the process. The general approach adopted by the Department of the Environment throughout the 1980s and beyond was to discourage local councils from 'interfering' in architects' individual design creativity and from getting involved generally in too much detail.

However, one is not at all sure that the Department of the Environment has been right to discourage local councils from having any involvement in detail. All too often, it is the detail, if carefully considered, which tilts the balance in favour of a generally sensitive and locally appropriate solution — whether it be the design of facades and fenestration, or the layout of parking spaces and hard or soft amenity areas. The DoE view (and that of many architects) is that the local authority's primary preoccupation in town planning terms should concern the more global matters, such as type and mixture of land uses, density of layout, amount of floorspace, amount of parking generation/provision, means of access, size and general massing of blocks to be placed on any particular site or area — even though this should not altogether preclude questions of detailed design being considered. Department of Environment advice (reflecting the joint consideration of the RIBA and the RTPI)[11] lays particular emphasis therefore on the need for development and/or design briefs[12] to be prepared in advance by the local authority for the initial or basic guidance of intending developers/applicants.

It is not practicable, of course, for a brief to be prepared for every building proposal that comes along — only the particularly sensitive, prominent or otherwise important sites or areas can be accorded this level of treatment, within the normal resource constraints of most, if not all, local authorities. It will always be a question of dealing with each site on its merits (via the development control system) according to its urban context and local circumstances, and within the financial and staffing resources available. Detailed design considerations will inevitably arise for the attention of the planners; i.e. questions such as positional relationship with other buildings (both in plan and elevation), external materials, general scale, roofscape aspects, fenestration in relation to facade design and appearance, parking layout, landscaping, etc. These will continue to be of legitimate general concern.

There may additionally be detailed architectural aspects which are also of valid concern to the local planners, in relation to the characteristics of specific nearby buildings, spaces or views, where precise amendments may be justified in order to protect or enhance the relationship of the new with the old. *Never, though, should the design consideration degenerate into a futile or blinkered debate about 'style' (that most dangerous and misleading of terms)* — rather, the consideration must be one of clear analysis of the development and design parameters, including that of context especially, followed by a constructive approach to the question of the form(s) in which the proposed development can be most appropriately provided on the site.

New versus Old

There is at any time a natural tendency for people to cling tightly to architectural forms from the past that are familiar, and to resist or reject anything that is unfamiliar. This is certainly understandable, and sometimes justifiable, particularly where the design problem consists of adding to, or inserting, a new building into an area of recognised specially cohesive scale and quality, where an 'alien' form of design would disrupt the whole area. In such cases, the presumption will normally be to reflect those existing characteristics of urban context in the new work or structure in a very direct way. In cases such as conservation areas, where the urban context is either entirely cohesive in scale and style or is a mélange of mixed but distinctive quality, new buildings may either replicate existing style(s) or (preferably) should be very sensitively designed to fit in satisfactorily as buildings of individual character. The essential thing is to maintain a balance whereby the existing overall character of the area is not prejudiced, but where there is room for manoeuvre as to new design forms.

The nature of the urban context must and will be the determining factor in the end, rather than preconceived notions of 'style'. As Capability Brown habitually said regarding 18th-century landscape design, 'Consult the genius of the place'. The same is just as true of urban design (i.e. where the 'contextual' area has genuine character or merit!). In this way, both appropriateness and spontaneity can be achieved, i.e. without stifling or unduly fettering the architect/designer. Rather than try to prescribe too specifically for each and every case in detailed design terms, the best approach to good design is probably that of the three principles historically laid down by Sir Henry Wotton, 'Commodity, firmness and delight', or in modern terminology, 'appropriateness to context, clarity of form relative to function, and an overall appeal to our innate sense of beauty' (quoted in Gloag 1950: 70, our emphasis). It is worth emphasising here that the use of local materials, or their equivalent within the local vernacular, is the single most telling and important factor, bar none, in ensuring a sympathetic building result.

Returning to the present-day personalities who take a central part in the current 'great debate' about modern architecture and urban design, the modernist architect Richard Rogers laid bare several of the main issues and arguments in his Foreword to Maxwell Hutchinson's book (1989) *The Prince of Wales – Right or Wrong – an Architect Replies*. He rightly identifies the seemingly perennial conflict between cost and culture, economics and aesthetics, as one of the main underlying problems contributing to the current climate of dilemma surrounding new building proposals. Rogers also rightly deplores the lack of public patronage and sense of adventure in this country (as opposed for instance to France), as far as new architectural commissions are concerned.

In this, he links the commercial and political institutions as the two real culprits for the frustration of the very debate which Prince Charles (in his well-known pronouncements on architecture) wishes to encourage. 'Ours is an age of business giants and cultural pygmies'. (ibid.: ix) How true. Blaming architects alone for cheap/shoddy (and unpopular?) architecture, Rogers rightly says, obscures the extent to which the large corporations, development and government are deeply implicated. His remarks would appear to apply mostly to the limited range of major prestige schemes with which members of the 'international circle' of architects are largely (if not wholly) concerned, but there is relevance also for the more 'local' types of scheme up and down the country. His thesis is that architects and architecture at any time must have simultaneous regard to the political, social, economic, fiscal and technological aspects of our rapidly developing society.[13]

He goes on to say that 'Architecture will no longer be a question of mass and volume, but of lightweight structures whose superimposed transparent layers will create form so that architecture will become dematerialised' (ibid.: xi).[14] Richard Rogers also states (ibid.: xii) that it is only [sic] through the development of the most modern ideas and techniques that we can solve both functionally and aesthetically the problems that confront us such as shelter for all, overcrowding, traffic, noise, smell and the general erosion of the beautiful qualities of land, sea and air. . . . it is a delusion to think that returning to a make-believe past can solve the global crisis. In fact, the danger we face is not of being too modern, but rather of not being modern enough.

Whilst Rogers is right in refuting the notion apparently adhered to by the Prince of Wales (amongst many other laypeople) that the conservationist approach will supply all the answers, *there is clearly a danger of rejection of all traditional forms and standards.*

Most, if not all, of the environmental concepts which he lists above are also good planning aims and objectives – one merely questions the use of the word 'only'. There is, however, no mention of reconciling and marrying the thrust of modern technology with the existing (often historic) urban structure. And the reference to 'a make-believe past' does not give one any added confidence that the justifiable fears and reservations of the public in Britain to so much change occurring to their urban (and rural) environment have been properly understood. As noted earlier, the danger is that of throwing the baby out with the bathwater, in terms of finding the best environmental and design solutions.

We shall return in detail to the all-important previous theme of design quality, and how this may be handled to best advantage by local authorities and others in reconciling the new with the existing. But first a general review of the statutory (and non-statutory) framework for overall planning administration at grass-roots level is necessary, since the nature and localised operation of the planning machinery must be the starting-point for examining how development proposals of virtually any type are dealt with, including the design aspects.

THE STATUTORY PLANNING FRAMEWORK: POLICY GUIDANCE

In very general terms, Town Planning in Britain is administered via two interlinked systems, informed by central government as to policy via a series of Planning Policy Guidance Notes (PPGs). The two related systems, which we shall examine in turn, are basically hierarchical, since the second depends to a large extent on the first. They are:

- The development plan system, and
- The development control system.

We shall examine the development plan system with reference to unitary development plans (UDPs), structure plans, local plans (now subsumed in UDPs and district-wide plans), plus various other forms of plan. This will then lead us, via planning briefs, etc., to look at how the development control system works, with extra detailed reference to conservation and preservation.

First, however, we set out a brief summary of the PPGs (current at 1996) that are mainly relevant to urban quality, since these encapsulate governmental policy attitudes, via the planning system in general, to the very questions with which we are chiefly concerned.

Environmental Principles

Central government guidance

Many government circulars and Planning Policy Guidance Notes (PPGs) deal with specific matters, but those which are of particular relevance to this section are as follows:

PPG1 (General Policy and Principles) states that 'Development Plans and guidance for particular areas and sites should provide applicants with clear indications of Planning Authorities' design expectations.'

PPG3 (Housing) emphasises the need for a high quality of design in all new housing developments producing buildings well designed for their purpose and surroundings.

PPG12 (Development Plans and Regional Planning Guidance) places emphasis, amongst other things, on the need to maintain the character of towns and the countryside and to safeguard and improve the amenity of residential districts. The PPG also regards energy conservation and global warming as issues that should be included in development plans.

PPG23 (Planning and Pollution Control) gives advice on the relationship between controls over development under planning law, on the one hand, and under pollution control legislation on the other. It is particularly relevant to industrial development, development of contaminated land, and waste treatment and disposal sites posing a potential for pollution. It also provides guidance on development proposals near such sites or land.

PPG24 (Planning and Noise) gives guidance on the use of planning powers to minimise the adverse impact of noise.

Circular 5/94 (Planning out Crime) gives advice about planning considerations relating to crime prevention.

Economic Activities

Central government guidance

The Planning Policy Guidance Notes that are particularly relevant to this topic are outlined below:

PPG4 (Industrial and Commercial Development and Small Firms) places emphasis on the need for development plans to take account of both the locational demands of business and wide environmental objectives. It points out that development plans provide a policy framework for economic development, weighing the importance of industrial and commercial development with maintaining and improving environmental quality.

PPG6 (Town Centres and Retail Development) stresses that a town's vitality and viability depend on more than retailing and that for a town to be successful there may be the need for refurbishment and modernisation. A balance must be secured between the need to sustain the vitality and viability of the town centres and those needs of customers best met through out-of-town development. Also, local authorities should ensure that disabled and elderly people as well as shoppers with prams and pushchairs have good access to shops and other facilities. It states that one of the roles of local plans is to assess the extent to which in-town and out-of-town retailing can complement each other in meeting the demands of the district.

In addition, a revised draft PPG on Town Centres and Retail Development was published for consultation in July 1995. The draft note, amongst other issues, places emphasis on a plan-led approach to promoting development in town centres and the promotion of mixed-use development and retention of key town centre uses. It also clarifies the three key tests for assessing retail development, which are: impact on vitality and viability of town centres; accessibility by a choice of means of transport; and impact on overall travel and car use.

PPG13 (Transport) says local plans should co-ordinate their policies for transport and other forms of development. It states that local plans should move towards achieving a better balance in employment and housing levels and should focus on activities attracting large numbers of trips in areas very close to major public transport facilities and in locations easily reached from local housing by public transport, cycle or walking (i.e. a sustainability theme).

PPG21 (Tourism) recognises the importance of tourism to the local economy. It sees tourism as a growth sector providing that it caters for the changing patterns of tourism and rising standards and expectations. It points out that planning should facilitate and encourage development of tourist provision, while tackling any adverse effects of existing tourist attractions and activity in a constructive and positive manner.

The Historic Environment

Central government guidance

Planning Policy Guidance Notes that are particularly relevant to this topic are as follows:

PPG15 (Planning and the Historic Environment) is concerned with all aspects of planning and the historic environment, including effective protection of listed buildings, conservation areas, transport and traffic arrangements in historic cores. It emphasises that there should be effective protection for all aspects of the historic environment.

PPG16 (Archaeology and Planning) looks at historic and archaeological sites and sets out the Secretary of State's policy on archaeological remains on land and how they should be preserved or recorded in both urban areas and in the countryside. It gives advice on the handling of archaeological remains and discoveries under the development plan and control systems.

Transport and Communications

Central government guidance

PPG13 (Highways Considerations in Development Control) updates and broadens advice in the light of new policy imperatives and the new status of development plans. The PPG advises authorities of ways to meet the commitments in the Government's Sustainable Development Strategy through influencing the rate of traffic growth and reducing the overall impact of transport. It accepts that forecast levels of traffic growth cannot be met in full, especially where this would be environmentally damaging. To maintain the effectiveness of the transport system, emphasis is placed on the need for policies to manage demand by promoting alternatives to the private car and enabling people to reach everyday destinations in urban and rural areas with less need to travel.

PPG7 Other government guidance deals with the environmental impact of transport policies in particular; road schemes in Areas of Outstanding Natural Beauty are considered in PPG7, plus 'brownfield' sites.

DEVELOPMENT PLANS

Those areas of Britain with single-tier local government, including Scotland, Wales, the major English cities and some other parts of England, are covered by unitary development plans (UDPs) which are comprehensive land-use plans. In those parts of England with two-tier local government, planning functions are split between county councils (which produce structure plans) and district councils (which produce local plans).

The purpose of structure plans[15] is basically to provide a range of planning policies at 'strategic' level for the overall area of each county in question. Except for certain strategic objectives (such as proposed new housing allocations for each district, or references to particular districts within, say, office or Green Belt policies), little specific guidance is given for local authority districts or parts of their areas, in terms of individual development sites. Site-specific comment is not generally made in structure plans, and any guidance in respect of urban design quality, conservation areas, listed buildings, etc. would normally be confined to very general references within 'environment' chapters or aspects of those plans.

In any case, structure plan policies will in future become even more broad and general in content, as they are replaced by a more succinct form of 'Strategic Statement' of former county-wide policies and objectives. In areas where new unitary local authorities have been created following the recent review of local government organisation, structure plans are being replaced by UDPs. Thus structure plans in themselves will have lessening relevance for urban design and conservation, though some county councils may continue to provide very useful advice to district councils on such matters.

Local Plans

These are produced by district councils for the whole of their areas, and thus are able to provide a more relevant policy context for urban design matters. A well-produced and comprehensive local plan will normally contain a fairly widely expressed chapter on environment, in which the local authority's general policies for defined areas of special environmental quality (including conservation areas) are set out.

There should be general policies relating to desirable standards of urban design (in terms of aspects such as scale, height and materials) for new development in different types of area, and for buildings of special architectural or historic interest (listed buildings). There should also be policies relating to areas needing environmental improvement (to whatever extent this is both necessary and feasible) and to additions and alterations to existing buildings in central areas and other sensitive areas (including significant detailed matters such as shop fronts, signs and street furniture) as well as for new buildings. Reference may legitimately be made to the local authority's own specific 'informal' policies for the control of detailed design for both commercial and residential extensions/alterations, and to some extent, infill development also. But these 'informal' policies (including replicated items such as leaflets and guidelines on design/conservation and trees/landscape) will need to be contained within appendices to the main body of the local plan, which will normally be concerned more directly with matters having strategic content or implications and resulting in policies carrying statutory weight after the local plan is formally approved by the Secretary of State.

Both structure plans and local plans have to go through a public inquiry form of examination, which gives the opportunity for full public participation, before the formal approval and adoption stages are reached. This is in order to ensure that all relevant aspects are thoroughly considered and tested to provide a statutory policy basis for development control. Inevitably, many local issues will have a distinct environmental context, where questions of urban quality and consequently the design aspects of new development will in turn be raised. Certainly the public eye will be focused firmly on these latter aspects, in addition to the policy principles involved in the development.

Other Plans

Reference should also be made to other parts of the development plan system, both past and present.

Local plans were intended to replace the old-style town maps and development plans which embodied a system of land use 'zoning' for clearly defined physical areas (making up a complete planning picture for the whole map area).

The development plan system, whilst clear and unambiguous, was felt to be too lacking in flexibility to accord with the modern-day needs of rapid and ubiquitous development pressures, and central government decreed that it should be replaced. After a brief flirtation in the 1970s with a new system named after PAG (the Planning Advisory Group), which was in reality only a permutation of the existing system but introducing such concepts as Action Areas, the local plan system was instead introduced.

This, as we have already discussed, was formulated on the basis of written general policies applying either throughout the plan area, or to certain defined sub-areas (e.g. a central area local plan for a sizeable town would normally have a defined Primary Office Area, in which special policies would apply to what in effect was the Central Business Area). The essential point, however, is that the local plan system was intended to replace the former development plan system, which only remained in force until approved and adopted local plans were in place. Indeed, the requirement for the mid-1990s was for all local plans to be quickly combined/subsumed in unitary development plans (or district plans), one for each unitary authority (or district) area, reflecting the reorganisation of local government during this recent period.

Two individual (but minority) features are those of Enterprise Zones and SPZs (Simplified Planning Zones). Enterprise Zones are designed by central government and enable rapid development, normally of a commercial or industrial type, to proceed virtually unfettered by planning and design controls. Simplified Planning Zones are areas set up by the local authority where planning restrictions are relaxed for certain types of development. Anyone can propose setting up an SPZ, including developers, and there is a right of appeal to the Secretary of State if the local authority turns the proposal down. Adopted schemes, however, last for 10 years. SPZs and Enterprise Zones need not unduly concern us in this essay upon the problems of reconciling new development with the existing urban scene, except to say that they perhaps afford a freedom of architectural and building design which might not always be appropriate elsewhere.

There are two features of the original town map/development plan system which, however, deserve special consideration, as still being of relevance. These are master plans and CDAs (Comprehensive Development Areas).

Master Plans

When the new local plan system was ushered in, it was felt that it could replace not only the old-style development plans, but also the old-style 'master plans'. The latter were a practical and useful means of laying out an overall physical pattern for an area that was clearly to be the subject of much change and development. They provided a cohesive picture of existing and proposed land uses, highways, open space etc., together with proposals for phasing. The needs of 'infrastructure' (i.e. services such as drainage, water, roads and even parking) could be fairly closely deduced. Moreover, the master plan could be supplemented by civic design studies and environmental proposals for special or centrally important areas. All this was rather pushed to one side in the all-pervading glow which accompanied the introduction of structure and local plans. Written policies would answer (almost) every need, it appeared.

It is now interesting to note that master plans have arrived firmly back on the scene (if indeed they were ever really absent) in order to provide a clear structure and pattern for areas subject to major development and/or widespread change. They were, of course, an integral feature of the New and Expanding Towns of the 1950s onwards, and of the rebuilding of war-damaged city centres in the same period (some of the latter, e.g. the Bull Ring at Birmingham, are now being rebuilt again). They were the necessary format for the major commercial schemes now being achieved in Docklands/Canary Wharf. With the necessary Department of Environment backing and structure plan/local plan policy context, they are also relevant ultimately to the redevelopment/rehabilitation of large areas of surplus or redundant land (such as the mental hospital 'clusters' in Health Authority ownership and closure programmes). Similarly, extensive areas of derelict industrial land adjoining certain 19th-century city centres can be redeveloped using phased master plan techniques applied to the city's local plans process.

In concert with the private sector, local authorities and urban development corporations still have the means at their disposal to create, control and, in part, design the pattern of major development.

CDAs

CDAs (Comprehensive Development Areas) should also be mentioned. During the 1960s and 1970s substantial areas of numerous towns and cities were run down and ripe for either complete or partial redevelopment, plus improvement. Where the local authority could control the land tenure situation, via its ownership of key sites and compulsory purchase of others, the CDA was the appropriate machinery. Whilst suitable and practicable in planning terms, the results on the ground were not always the best in terms of architectural design. The most common problem was that areas that were all broadly developed at the same time ended up with a monotonous and soulless appearance, accentuated (or perhaps caused) by the rather sterile 'drawing-board' architecture so typical of the 1960s. Design schemes which looked fresh and interesting in drawing or model form, or even when newly built, soon took on a character similar to that of the reconstruction of Le Havre, with flat-topped blockish buildings, gridlike fenestration, and stained and weathered concrete. The 'Grand Project' can have a stultifying effect (and one that quickly dates) and furthermore is out of sympathy with the English tradition of small scale and intimate character.

But the concept of integrated planning areas, as against the CDA concept, is not entirely without validity, where a whole area of, say, a town centre needs to be planned on the basis of a cohesive (but not single!) land use and highway/access/parking pattern. One of the essential ingredients would appear to be that of the local planning authority treating the area on an incremental, phased basis, with individual developers, architects and developers each responsible for their own part of the total area. A desirable further ingredient is for the planning authority to have some degree of overall design concept for the area, at a reasonably early stage. In this way, the range of development rising eventually from the 'jigsaw' not only works to a cohesive plan, but also exhibits an interesting variety of form. An increasingly necessary element for local authorities to utilise nowadays is the planning/development brief, which we have already discussed above, together with basic questions of design control, following the remarks on local authority resources at the beginning of this section concerning the statutory planning framework.

Development and Planning Briefs

Distinction needs to be made in fact between the planning brief and the development brief. Both of these stand halfway, as it were, between the statutory local plans and the statutory development control system, but only carry formal weight if they are officially approved or adopted by the local authority as formal policy documents for the control/development of the specific areas in question.

The planning brief normally sets down planning guidance for the whole of what may in fact be a very mixed area, in which various and differing development needs and opportunities exist. It will directly stem from the local (or unitary) plan policies covering and including that area. As such, it will be principally a set of planning policy considerations affecting and guiding any development taking place within that area, but will not normally lay down development guidance for specific sites or sub-areas. That will be the task of the development brief.

The development brief, as discussed above, will be site specific, and may refer either to one individual site or to a number of sites within the overall area which was the subject of the planning brief. It will thus be even more closely related to the development control process than the planning brief, and its contents will incorporate such aspects as range of acceptable land uses linked to appropriate densities, physical layout considerations including means of access (primary and secondary), plus broad location and extent of buildings, floorspace quantum/ land use, etc. It will also relate/refer to the planning brief, or may in certain cases be combined with it (i.e. merged, in the case of individual sites, or as a joint document in the case of a number of sites within a common area).

A third potential category of local authority advisory document or 'brief' is the concept of Design Guidance Codes, or design guidelines. Traditionally, design advice was only administered informally via negotiations between a council's officers and the developer's architects, as part of the development control process (either before or after a formal planning application is submitted) However, the Secretary of State for the Environment announced in a keynote speech (entitled 'The Role of Government in Architecture') at the Royal Fine Art Commission (RFAC) on 13 November 1990 that the government was considering the introduction of local planning authority design guidelines, which he personally advocated. He accepted 'that the public interest requires protection through public action, and that a degree of objective standard-setting in urban design is both possible and desirable'.

The government's proposals formally to allow local authorities to introduce their own design guidelines (hitherto only practised by councils on an unofficial basis) thus recognised at last that local authorities do have a legitimate role in the building design process 'in matters such as scale, density, height, massing, layout, landscaping and access'. The Royal Institute of British Architects, whilst not publishing a supportive statement, was apparently able to accept the government's suggested checklist of essentially broad matters and *thus the concept of an 'environmental audit'* ('style' was however excluded).

The Secretary of State for the Environment, in his RFAC speech, stressed the need to recognise that 'good design is a good investment', and said that the study that the government would now set in motion would (*inter alia*) look at arrangements adopted by other major institutions and companies to ensure that buildings resulted from the right brief and the right designer. He also emphasised, however, that in according a place for design guidelines in the planning system, a careful step-by-step approach would need to be adopted in their production to make sure they were not 'absolutist or prescriptive', or did not simply encourage 'pale imitation'. He said that there were four long-established proposals on good design that he intended to implement. First, that a building is meant to be used, and that it should therefore be possible to read it 'like an open book'. Second, that the design and scale of a building should reflect its use (similar to the first point). Third, that it should 'line well' with its surroundings, even if it does not slavishly observe established styles. Lastly, that it should reflect the community that lives there. He went on to emphasise that a 'battle of the styles' in every council chamber must be avoided, suggested that we could learn from the past, when cities benefited from building codes, and pointed to the planning practices of certain US cities, which have design guidelines that developers have to take into account. These guidelines were not regulatory, so that there was still scope for the 'beautifully eccentric or the brilliantly unexpected'. The Secretary of

State suggested that such guidance could provide a developer, in advance, with a clear indication of the authority's expectations, and act as a kind of checklist — but one that 'stimulated not stymied'. He then proposed that a few local planning authorities could produce model guidelines to provide common ground to be followed by a Design Circular in the forthcoming year, plus a Planning Policy Guidance Note.

Later, the Secretary of State, addressing the Policy Studies Institute, said that he wanted to see public standards 'so high that no-one will seriously believe that the private sector should be an automatic choice for those who have the resources to opt for it'. In relation to the subject of this thesis, it is worth highlighting his related views that:

- civic design skills have been neglected of late;
- 'good civic design is something that an age of individualism needs to keep constantly in mind';
- planners, in particular, must be sure to bring adequate design expertise to their key mediating and advisory role in the planning process; and
- if public design standards are to be maintained (and enhanced where necessary), resources clearly will have to be committed to the public sector to train and employ the necessary urban design staff.

It is an open question [=training] however, as to whether the necessary resources have in fact been made available for this staff, when local authorities have been left painfully short of funding in general.

DEVELOPMENT CONTROL

The principal bulwark of the statutory planning system in day-to-day terms, and complementary to development plans, is of course the development control system — the 'sharp end' of planning, where decisions on proposals have to be made. In so far as urban design and quality are affected, we are principally concerned with the following broad elements:

- Application format, and supporting information material, submitted for planning and/or listed building consent (also, under a separate system, Building Regulations approval);
- Consultations, negotiations and assessment at officer level (normally including any design or related negotiations prior to formal application stage);
- Formal consideration of application and decision at either officer or committee level (only 'delegated matters' are decided at officer level) on the basis of planning officer assessment and recommendation. Inspection and enforcement matters/decisions;
- Consideration/hearing of appeals where consent is refused and the applicant's dispute is referred to the Secretary of State for decision. (The Secretary of State may also 'call in' an application for decision by himself in placing the local authority's formal 'determination' of the proposals.) Conditions of approval are imposed by the Secretary of State, where applicable. Award of costs are made where applicable/merited — in either direction;

It will be noted that both the assessment of the proposals, and the procedures thereto, leading to formal decision, are closely interlinked throughout the development control process. (The question of timetabling relative to the processing of the application also underlies all these elements — more of this aspect later, together with that of overall economics and cost of proposals.)

The range of participants who take a hand in the planning and 'urban design' process is extremely diverse, but may be broadly summarised as follows:

1 The applicants (actual or prospective) who may be either private citizens/householders,

private developers, or sometimes public bodies. They may or may not be the owners of the land/site in question.

2 The applicants' agents, who are frequently land surveyors/architects/ designers/commercial advisers/engineers and also chartered town planners nowadays.

3 Local objectors and other persons (e.g. neighbours) with an active and/or pecuniary interest in the proposals, whom the local planning authority is under varying degrees of obligation to consult.

4 Consultees additional to those in (3) above; these will include the Highways Authority, principally where either highway control/design, layout, access or parking issues relative to classified roads, or where strategic development plan/local plan policy issues are involved. Local residents' associations and amenity societies will normally be regular planning consultees, as will town and parish councils.

5 Historic buildings bodies in the case of proposals involving change to listed buildings (and other buildings of historic interest forming part of a group, more particularly in conservation areas). The chief point of reference for listed buildings is English Heritage (which grew out of and in effect replaced the Department of Environment's Historic Buildings Division), but specialist bodies who must also be consulted where schemes involve demolition of historic buildings include SPAB (the Society for the Protection of Ancient Buildings), the Georgian Group, the Victorian Society, etc. (a total of six 'national bodies' in all).

6 Specialist consultants: planning, commercial and architectural consultants are frequently employed by private developers and also occasionally by local planning authorities, where proposals concerning important listed buildings or of an otherwise major/complex/ or sensitive nature or context are involved. (The Royal Fine Art Commission may also at times require to be consulted.)

7 The local authority, i.e. the body to whom the application is submitted, and who normally 'determines' it (viz. comes to a formal decision upon it, which is of statutory effect unless appealed against within a prescribed period of 6 months). The local authority operates via its officer structure (Chief Executive, Chief Planning Officer and staff, and consultee departments) and its committee/member structure, with its full council normally deciding or commenting only upon specific matters referred to it by the 'service committee', i.e. the planning committee in this context.

It should be noted that some Planning Authorities have their own specialist urban design and conservation staff (as discussed earlier in this appraisal of the local government system, and its provisions for design advice and control), and the need for this appears to be increasingly recognised though there are not sufficient specialist personnel of this nature to spread over all local planning authorities. Whatever the specialist staffing provision, however, the development control support staff need at least to be able to recognise a design issue when one arises, and either to deal with it themselves (if fairly straightforward) or otherwise to refer it to a more senior/specialised level (where of more complex or sensitive nature/context).

The Chief Planning Officer (or Director of Planning Services) should also possess an adequate degree of design sensibility (if not training and/or experience) to be able to formulate advice both to the applicant and to the planning committee upon urban design, conservation or environmental improvement matters. There has to be in each local authority area someone of perception and vision at the planning apex, who can form, retain and drive forward a concept of beneficial urban change together with a sensibility for detailed urban design and control of development.

8 Finally, there is the Department of Environment, which is the controlling government body for the operation of the planning appeal system, and which also monitors quarterly the development control performance of local planning authorities, in terms of the numbers/proportion of planning applications 'determined' within the 8-week statutory period.

Note: the 2-month statutory determination period has existed unchanged since 1947, despite various reviews, and despite the ever-changing picture of development pressures as against council staffing resources – or perhaps because of the fluctuating picture. In theory, all applications should be determined within this period – in practice, it is impossible for more than about 50–60% on average to be so determined, owing nowadays to the major size and/or the complexity/controversial nature of many development proposals. The number of applications which are dealt with inside this period mainly comprise the relatively simple or straightforward 'householder' applications, dealt with at officer level. Where trunk roads, involving Department of Transport (DTp) consultation are involved, a 3-month period for determination will be applicable, but in practice up to 6 months for decision is not uncommon for major development schemes of any significance. Some authorities manage to achieve a high proportion of all applications being dealt with in under 2 months, but this is often dependent on 'reserved matters' (i.e. details following grant of outline consent) being treated as separate decisions statistically.

The relevance of all this for design control by councils is that some developers complain that extra negotiations in the form of design advice mean extra time and cost added on to the overall timescale for decision of the application proposals – a very real problem in economic and programming terms for most developers. Complaints regarding the total time needed for a planning decision ran for a while like a river in full spate during the early 1980s, when at central government level there was much dark talk of 'jobs locked up in planners' filing cabinets'. This problem is now regularly circumvented by the practice of 'twin tracking', i.e. that of developers putting in two near-identical applications simultaneously and then lodging in advance an appeal on one of them on the grounds of 'failure of the local authority to issue/make known its decision within the statutory 8-week period' – a move which, though just about legitimate technically, is not exactly calculated to establish the best possible basis for negotiations (and approval) on the part of the local authority, including the matter of design negotiations. It also clogs up the DoE's appeals system, as well as the local authority's development control system, and the government has now called this dubious practice into question.

Getting development proposals dealt with sensibly and even-handedly via the planning 'minefield' in Britain becomes ever more frequently a frustrating, uncomfortable and even damaging process, resolved increasingly for good or ill on appeal. At peak periods, planning departments (with their usually slender staff resources) often perform near-miracles to keep on top of the work influx. At appeal, in the public inquiry situation, local authority planners may seem, on the face of it, to be out-gunned by developers, who are able to produce large teams of expert witnesses and weighty piles of evidence (sometimes in bulk standardised form, for the 'appeals circuit'). But the rate of dismissal of planning appeals nationally still says something for the integrity and efficiency of the UK town planning administration.[16]

In all this, it is even more remarkable that a high quality of development is obtained for the local environment, via attention to design. *The saving grace is perhaps that more developers now realise that good attractive design not only helps to secure planning permission, but that it is also a good investment.* This is also appreciated by well-informed, progressive local councils.

CONSERVATION AND PRESERVATION

At a time of widespread pressure for rapid changes of varying types to the urban environment, it is perhaps our historic areas and buildings which are most at risk. The demands for increased mobility and access, modern standards of accommodation, new building technology and new scales and styles of development, linked in turn to changing demographic patterns, new levels and distribution of wealth, and changing use requirements, all combine to exert a profound effect upon the built environment and its residual features from the past.

The effect of this currently is to cause a revaluation of attitudes by the public to both development and conservation, and a revised approach by local authorities and central government to these matters. It therefore seems appropriate to consider current changes and trends in urban conservation and building preservation, in the light of fresh circumstances and prevailing legislation in Britain. In this, however, it should be emphasised that urban quality is something that needs to be striven for not only in areas of historic sensitivity but also generally in our towns and cities.[17]

Origins, Concepts and Differences of Purpose

Conservation areas as a concept were originally created by the 1967 Civic Amenities Act, under Duncan Sandys as the Minister for Town and Country Planning. This piece of legislation differed fundamentally from existing powers of protection of the built urban heritage in that it recognised and sought to protect the overall character of historic areas, and implied a conscious 'civic design' approach to any proposed changes within such areas. Previously to this, the only basis for safeguarding the historic built environment was via the preservation and protection of individual listed buildings, though the Statutory List of Buildings of Special Architectural or Historic Interest did extend to the listing *per se* of significant groups of historic buildings.[18]

It is important to draw a distinction at the outset between the purpose of listed buildings and conservation areas as concepts, with their respective legislation. The essential criterion for listed buildings, as already indicated, is that they should possess 'special architectural or historic interest', which should be preserved against unsympathetic and unsuitable alteration, conversion or other change which might adversely affect the building's character (or setting), or against partial or full demolition. The accent has traditionally been upon 'preservation' of listed buildings, in terms of their form, appearance and/or setting – whereas the concept of conservation areas has always been rather more liberal, i.e. allowing for a degree of 'organic change' whilst retaining the basic characteristics of the area. (Thus 'conservation' does not equal 'preservation', though there are many who confuse the distinction, or who for various reasons are unwilling to accept this.) The 1971 Planning Act introduced powers of control over demolition within conservation areas, and it is now necessary to make a listed building application for any such demolition proposals.

TOWN PLANNING FRAMEWORK

Buildings and Conservation Areas) Act of 1990, as 'areas of special architectural or historic interest, the character or appearance of which it is desirable to preserve or enhance', whilst the present conservation areas concept (as amended by successive pieces of legislation) is contained within Sections 69 to 75 of that Act. The original concept would normally be centred upon a significant presence or grouping of historic buildings, often with some additional context of interest or distinctiveness, and this conception is still basically as envisaged by the present Act. The presence of particular open spaces, or certain physical views in addition to the essential 'nucleus' or quantum of historic buildings (or other listed features, e.g. walls or enclosures), may all go towards making up the conservation area characteristics, which should by and large be retained, but in which context sensitively considered and well-designed changes can take place. Within conservation areas, there is an expectation that unsightly or unneighbourly development will in time be replaced by new schemes that enhance the character and appearance of the area.

Powers and Procedures of Local Authorities

There are important differences also in local authorities' powers and procedures, as between listed buildings and conservation areas. The Secretary of State through English Heritage has the responsibility for 'listing' buildings. This sometimes comes about via requests from the local authorities to list buildings that they consider worthy of protection, but sometimes the Secretary of State may do so (or may review and revise the existing lists) on his own initiative. He may also be lobbied by special interest groups (amenity societies and the like) or even by individuals to list buildings, either via a local authority or directly. In other cases he may be requested by planning applicants and/or owners to 'de-list' certain buildings by

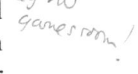

Figure 8 Preservation or Conservation? Bristol.
The commercial heart of Bristol, a thriving centre through medieval, Georgian and Victorian times, now pedestrianised and largely for disposal. The impressive commercial and religious architecture is redundant.

This high urban quality, realised over centuries is unsuited to the motor car, too small in its accommodations, and very expensive to maintain: a repeating story in European cities today. A precious environment will not be preserved if its economic use cannot conserve its real value to local society.
Source: G. Power by permission of J. Arthur Dixon

Similarly, it was the 1974 Act rather than the Civic Amenities Act of 1967 which placed statutory obligations upon local planning authorities to prepare proposals to preserve and enhance conservation areas, in addition to creating or 'designating' such areas. The statutory definition of conservation areas is now contained in paragraph 69 of the Planning (Listed removing them from the Statutory List. Listing of buildings by special request of individual bodies or persons is known as 'spotlisting'.

Listed buildings are now classified as follows:-

- **Grade I**   Buildings of 'exceptional interest' (e.g. of or equivalent to a national level of architectural or historic significance);
- **Grade II***   Buildings of 'particular importance and perhaps containing outstanding features';
- **Grade II**   Buildings of 'special interest which warrant every effort being made to preserve them'.

It is estimated that Grade I buildings comprise only about 1% of the total, and that Grade II* buildings number some 20,000; the vast majority of listed buildings and other structures are Grade II. The previous category of Grade III was withdrawn, concurrent with re-listing generally, and the majority of its buildings were transferred to Grade II.

Statutory Listings, and the process of listing, are regularly used by local planning authorities to halt the demolition or prevent the unsympathetic alteration of a building. The planning authority exceptionally may issue a Building Preservation Notice to protect from alteration or demolition a building which they have requested the Secretary of State to list. The Building Preservation Notice is immediately effective, and remains in force for 6 months, or until the Secretary of State decides whether or not to confirm the listing. (There is no right of appeal against listing, but an applicant may of course appeal against refusal of listed building consent. There is also provision for obtaining, via a planning application for the future use and form of the property, a certificate of immunity from listing, valid for 5 years if confirmed by the Secretary of State, but this does not necessarily lead to permanent immunity.)

The essential difference procedurally between listed building and conservation area statutory provisions is that the Secretary of State for the Environment decides upon all 'listing' cases, and controls important LBC (listed building consent) decisions. He must be notified by the local authority, for example, of its intention to approve any LBC applications for demolition, and must respond within 28 days, and may also 'call in' any major or controversial LBC proposals for his decision via a public inquiry. The designation of conservation areas, on the other hand, is in the remit of the local authority – a virtually unique circumstance where the definition and protection of the historic environment is concerned. The local authority may actually designate a conservation area within 1 or 2 days, if the urgency is warranted by a threat of demolition or similar.

The designation process to be implemented by the local planning authority is not particularly complex: the authority normally prepares a 'statement of intent' plus a map defining the proposed conservation area (at a suitable scale to identify individual features and properties, yet covering the whole area), with a set of property references, plus notices in the *London Gazette* and a local newspaper. The council will usually declare publicly its intention to designate the conservation area, before making a formal designatory decision, but in fact is under no statutory obligation to consult anyone up to this stage, and there is no formal right of appeal against designation or any means or form of public inquiry relative to the question of designation. The 1990 Act also places upon local authorities the duty to review their overall areas, to see whether any existing conservation areas should be altered or extended, or further areas designated, setting out the formal process.[19] It is important to note that, at the time of designation, a document should also be prepared by the authority defining and making available for any further reference what the main characteristics of the proposed conservation area(s) are, and preferably also setting out policies or guidelines for new development and any other changes envisaged, together with any key features or buildings (listed or unlisted) to be retained.

Protection and Enhancement: Resources and Practicalities

As already stated, the 1974 Act required local authorities to prepare proposals to preserve and enhance (as appropriate) conservation areas within their localities, and this is still the position. However, the financial resources (and indeed the freedom) of local authorities to prepare and carry out physical enhancement schemes were considerably greater in the 1970s than in recent times owing to increasing central government restrictions on local government activity and spending. Also, the number of designated conservation areas in existence in the 1970s was considerably less (there were nationally a little over 3,000 conservation areas in 1975, compared with an early 1990s level approaching an estimated 7,000).

Public Attitudes towards Development and Conservation

The escalating pressures for development so evident throughout the 1980s levelled off and relented in the early 1990s owing to the economic recession, but an increasing antipathy had by then been bred in the public's mind towards the effects of unbridled private development, both as affecting the more central historic parts of the environment and in respect of other areas (such as residential).

This has brought about a defensive and protectionist attitude generally, which manifested itself nationally in backlashes such as 'NIMBY' (Not In My Backyard) and was given further impetus by Prince Charles's public crusade against brutish or otherwise inappropriate forms of Modernist architecture, and by his concomitant espousal of conservationism and a widespread return to past styles of building. The public debate over all this is still in full swing, and shows little sign of abating. It goes much wider than mere 'conservationism'; in fact, it is by no means proven that the general public are fully and wholeheartedly in favour of conservation and the protection of historic buildings.[20]

In some quarters, notably among the more articulate and well-informed sectors of the population, this has given rise to a somewhat 'legalistic' approach to conservation, as well as a generally 'protectionist' one. Local planning authorities have for some time now been largely thrown back by economic stringencies and practicalities to their traditionally 'negative' development control approach of maintaining and defending the quality of the environment via local control of individual planning applications. They have at best sought to ensure that building proposals have fitted in as satisfactorily as possible with neighbouring development, rather than requiring them to raise the standard or appearance of their surrounding urban context. For a long time, this has been the planning authorities' approach to development both outside and within conservation areas, except perhaps for seeking a more sensitive level of design within those designated areas. This approach has also gone hand in hand with the increasing difficulty of finding the means for carrying out enhancement schemes in conservation areas, despite public pressures and expectations in favour of such schemes and despite the duty placed upon local authorities under Section 71 of the 1990 Planning Act to 'publish proposals for . . . the enhancement of . . . conservation areas' (see also 'Enhancement Opportunities', p. 48–9).

In 1989 an extra weapon fell unexpectedly into the hands of the public preservation and conservation activists. A decision of the High Court in Steinberg and Sykes v. Secretary of State for the Environment clarified the duty of the local planning authority under Section 277(8) of the 1971 Act, now subsumed in Section 70 of the Planning (Listed Buildings and

Conservation Areas) Act of 1990. This legislation provides that 'where any area is . . . designated as a Conservation Area, special attention shall be paid to the desirability of preserving or enhancing its character or appearance'. The judge ruled that 'harm is one thing; preservation or enhancement is another. . . . The concept of avoiding harm is essentially negative. The underlying purpose of Section 277(8) seems to me to be positive.' The decision was generally seen not only as giving planning authorities and amenity or conservation bodies more control and influence in conservation areas, but also as placing a greater degree of obligation (if not duty) upon local authorities towards (a) preparing and implementing schemes of improvement for conservation areas, as necessary, and (b) requiring and ensuring that individual development proposals and planning applications in conservation areas should positively enhance their sites and surroundings – i.e. that merely securing that the proposals 'fitted in' was not enough.

However, in the same year, Sir Graham Eyre pronounced in South West Regional Health Authority v. Secretary of State that 'the subsection does not set out a test, nor does it support the proposition that . . . development proposals . . . must themselves in every case preserve or enhance the character or appearance of the Conservation Area', and 'The Steinberg case does not rule out the investigation of harm in the context of paying special attention to the desirability of preservation or enhancement.'

Local authorities up and down the country breathed again, but the Steinberg case had by then been taken very much on board by local amenity societies, and still continues to be invoked by them from time to time. Planning departments, it must be said, conscientiously resist overdevelopment generally, and normally try to ensure that new building proposals in old established areas are designed to relate as satisfactorily as possible to their context (making a positive contribution wherever appropriate), but it is clearly unrealistic and artificial to expect every such proposal to provide a major enhancement. Steinberg has since been overtaken by other legal rulings and cannot be relied upon.

Nevertheless, Steinberg not only has provided the means towards a more purely conservationist approach, but much more significantly is seen as an extra means of resisting and frustrating any unwanted development – and that can cover a whole multitude of sins!

Benefits of Conservation in Practice

Despite the daunting restrictions placed on local authorities nowadays as to any form of expenditure on area improvement, some opportunities and means still exist for enhancement (though at a price) as well as actual protection of the historic environment to take place. In examining these, we should perhaps start by considering the position concerning individual listed buildings, before looking at the wider picture.

There are of course *specific obligations upon owners of listed buildings to maintain and preserve them*, in addition to being subject to a more than average standard of requirements if they wish to alter them, add to them, or change their use in any way. However, a degree of flexibility of land use (e.g. office consents), coupled with relaxation of building regulations standards, may be obtained in order to secure the continued existence of such buildings. The fact that they are often found to be not fully compatible with present-day domestic, space or health standards is not necessarily a condemnation of those old buildings. After all, the buildings were there before somebody invented the standards – and in most cases they still have much to offer, both aesthetically and in practical terms, so a sympathetic approach to their retention and extended use is appropriate on all sides.

Owners of listed buildings thus have to keep

TOWN PLANNING FRAMEWORK 47

Figure 9 Bristol/Lincoln: Examples of preservation.

Fine buildings from the past can outlive their original purpose in less than a generation and mere preservation cannot ensure continuing commercial value. Conservation, on the other hand, is often able to secure another lifetime of irreplaceable value, by the sensitive incorporation of fine architecture into more 'neutral' but economically viable city centre space.

The Edward Everard Building in Bristol (left) forms a fine entrance to a Natwest Bank Regional Office, in a much-needed commercial rebirth of the city centre.

The medieval building in Lincoln (below) owes its state of intactness and wellbeing as much to its prime position near the Cathedral as to its individual use and city preservation policies. It houses the tourist information centre, and as such is ideally placed in the historic city core.

Sources: Bristol photo, Gordon Power
Lincoln photo, Michael Parfect

them in a reasonable state of repair, but in some cases (particularly if this is genuinely not fully within the means of the owner) grants may be available from the appropriate local authority for repairs. Grade I buildings will attract a higher level of grant than Grade II cases, and the more important examples may well attract additional grants from other sources (e.g. from English Heritage) though normally for only a part of the total cost, and depending on circumstances. Alterations to listed buildings which have received listed building consent and some restoration works are exempted from VAT (value added tax), but day-to-day repairs are not. Notably, though, the planning rules for listed buildings apply also to all buildings within their curtilage where these have been erected before 1948, including structures in gardens or courtyards, and other buildings in the same group.

Generally, however, in a climate of very high building costs, the virtues of retention and refurbishment of older buildings are becoming increasingly recognised. This brings additional benefits in its train, such as energy savings in manufacture and use of materials compared to the demolition, structural and work-energy input required by redevelopment schemes. Also, refurbishment is normally more 'neighbour and environment friendly', being less disruptive in activity and noise impact terms than 'new build', as well as more easily assimilated within the urban context in aesthetic or design terms.

In those difficult cases where the owner declines to look after the listed building as he should, the local planning authority can put pressure upon him or her to do so to the extent of compulsorily purchasing it in appropriate instances (e.g. where the owner is actually disfiguring the building by unsuitable alterations – such as covering a Georgian brick or stone building with pebble-dash), or by requiring the demolition of unauthorised extensions that positively detract from the character of the listed building.

Local authorities are normally reluctant, however, to take action to redress a situation where this involves financial expenditure on their part which may not be recoverable in return. Nevertheless where the legislation on listed buildings and conservation has teeth, it is frequently used.

Enhancement Opportunities

Turning to the wider picture, it may be seen that a policy of active conservation and area enhancement where possible has real and positive benefits. Apart from recognising the truth of the dictum that 'a town without old buildings is like a man without a memory' (attributed to Konrad Smigielski, the City Planning Officer of Leicester in the 1960s, when the conservation movement was putting down its first roots), a major and obvious beneficiary of conservation is tourism. Tourism is nowadays a multi-million pound industry in Britain, as elsewhere, and was recognised as such by the late 1960s, and certainly by the mid to late 1970s, when expenditure by overseas visitors increased from £359 million in 1969 to £2,179 million in 1977, i.e. an increase of £1,820 million over 8 years, or nearly £230 million per year.

The overall picture regarding tourism and its relationship to the preservation of the urban heritage was examined in depth at the end of the 1970s in the booklet *Preservation Pays*, written by Marcus Binney and Max Hanna, for SAVE Britain's Heritage. Twenty years later, the financial returns and benefits nationally of tourism relative to conservation and preservation can only be commensurately much higher. It is outside the scope of this deliberation to go into the overall position and the arguments in detail, but the implications for all-round benefits to the business life of the community in general are surely evident.

Similarly, there are clear benefits to the community to be gained by linking planning objectives to area enhancement. The most obvious example is probably that of 'planning gain', whereby the local authority secures certain enhancements (in the form of essential local improvements) to the environment, concomitant with (and in effect as a condition of) the grant of planning permission for development proposals of a substantial nature on a prominent or contentious site and/or where planning policy complexities warrant a special approach, or even concessions towards the developer. This type of approach may indeed be contentious, as it could in some cases smack of 'selling planning consents', but it is sometimes necessary in order to resolve a deadlock or ease a situation. *It has clear connotations with the normal conservation area situation*, where every potential development site poses sensitive issues, and real opportunities for physical improvements need to be taken, but where the necessary funding would not normally be available, but for the 'spin-off' funding benefits arising from substantial development changes.

Another type of enhancement linked to planning aims is where, for example, a new road or traffic system brings about the opportunity for a pedestrianised (or part-pedestrianised) area to be created. The absence of traffic movement (apart from essential servicing and special access and parking needs) can in turn benefit the area economically by making it more safe and attractive to use, which then more strongly justifies and encourages the provision of physical improvements (such as better street furniture, lighting and paving, together with ancillary planting). Shoppers, tourists and residents, together with shopkeepers and traders, all benefit, and more money can be found to underwrite and further improve the operation. *The importance, however, of adequate traffic arrangements, including servicing access and parking provisions, cannot be overemphasised in such proposals.*[21] One of the earliest such instances in Britain is reputed to have been London Street, Norwich, where a long, narrow, winding and uphill shopping street was rescued from oppressive traffic conditions and consequent decline by a watermain burst necessitating complete street closure. The City Planning Officer (Alfred Wood) saw his chance and speedily converted the street into two pedestrian precincts (upper and lower); London Street never looked back thereafter. (Salisbury and Southend-on-Sea also have competing claims to be the earliest such case.)

Unlisted Buildings in Conservation Areas: the Need for Greater Control

A source of particular concern in conservation areas is the damaging cumulative effect of relatively minor but numerous changes to unlisted buildings, which thus do not require planning consent; this would certainly be otherwise in the case of listed buildings. Fortunately, all demolition proposals in conservation areas now require express consent, but a plethora of minor physical changes not needing consent can detract almost as seriously from the quality of the historic urban scene in terms of overall degradation.

This unhappy situation was examined in depth by the English Historic Towns Forum, at a seminar hosted by the Hove Borough Council in June 1991. A whole range of problems was addressed, ranging from plastic window frames replacing original-style timber fenestration and original details of facades being obliterated, to rooflines being wrecked by large dormer bulkheads, to the proliferation of garish painting schemes, and front gardens being hard-surfaced and given over to parking. 'Cheap and cheerful' can easily mean vulgar, disruptive and generally inappropriate in the conservation area scene. Add to this the detrimental effects of excessive advertisements, and flyposting to boot, and normally attractive and sensitive areas can quickly go downhill in both appearance and value.

There exists, of course, the 'Article 4 Direction' power of a local authority to remove 'permitted development rights' in conservation areas, as it sees fit, together with the normal enforcement powers, *but many planners now feel that a general widening of detailed control over unlisted buildings and minor changes in conservation areas is requisite.* 'Codes of Control' may be introduced as a further measure, relative to the removal or restriction of p.d. (permitted development) rights, as a less formal means of control than conservation area powers or Article 4 Directions.

It goes without saying that the most sensitive possible standards of design should be applied to all new development in historic areas, including alterations and extensions in conservation areas, and particularly to listed buildings, whatever their location or context. This of course is the vital clue; the nature and characteristics of the urban context of individual buildings or sites will in most cases itself provide the necessary additional design guidance. The urban or architectural context must always be taken into account, and indeed the proposals submitted for planning consent should always:

1 show accurately in both plan and elevational form the essential relationship between the new and existing buildings (the Royal Fine Art Commission, in their 1990 publication *Planning for Beauty* underlines this need); and
2 provide both written and drawn visual evidence justifying the purpose, needs, scale and design of the new development, preferably with reference to spatial effects as well as physical and material effects on the townscape or environment.

The Way Ahead

Clearly, there is no universal panacea for dealing with the problems of conservation and preservation, or indeed for eliminating the degree of confusion that frequently exists between the two in the public's mind. There are, however, a few general points which may be quoted as indicators of a broadly appropriate way in which to proceed, as follows.

1 In line with the need to accept that our whole way of urban life is rapidly changing, there has to be a general policy of allowing controlled but sympathetic change in conservation areas rather than an attempt to misuse the planning framework to stifle or frustrate most change, in a NIMBY-like fashion.
2 The reality of the sharp imbalance between the growing (and arguably now sufficient) number of designated conservation areas and the lessening ability of local authorities to respond other than by applying a firm (but hopefully sensitive) development control approach should be recognised by those who clamour for further enhancement measures to be taken at every turn, wherever any degree of controversy arises.
3 The concomitant reality of the acute restrictions upon local authorities, financial and staffing resources should also be recognised, with particular reference to the shortage of trained conservation and design staff (as against the number of local planning authorities in the country) and the ever-tightening squeeze upon money available for physical improvement projects. (The paramount requirement is for staff to be available at least to deal adequately with urgent cases of threats to listed buildings, either via physical decay or dereliction, or via totally inappropriate proposals for conversion, extension or even demolition.)

With respect to this last point, the important ancillary role of the amenity societies should not be overlooked or discounted. They are frequently able to provide well-informed comment on local situations, and can strengthen the planners' arm by drawing their attention to, or forewarning them of, actions or proposals that would be environmentally detrimental. Their contribution should always be carefully considered and heeded by the professionals, as well as by the politicians generally, as long as it does not become 'reactionary', i.e. resistant to change in principle.

It is largely, however, through an open balance of protectiveness and enterprise, control and innovation that the very charm and interest which so characterise the best of our historic urban heritage exist. An intelligent and understanding approach from all sides will always be the prime factor in successful environmental preservation for our continued use and enjoyment.

THE DEVELOPMENT PROCESS IN CONTEXT

Having broadly examined the conduct of the two main 'strands' of the town planning system, i.e. development plans and development control, we now need to look in more detail at how the development control system regulates and affects the actual process of development and building, in terms of approvals and implementation. The relevance of all this to the provision of design advice, and acceptable environmental standards, continues to be the underlying theme.

The developer/applicants' initial consideration is most likely to be that of timetabling owing to the direct effect of the period needed for an approval/refusal decision (and any appeal) upon his financial position, via the loan charges and the development costs – basically the site and building costs, both of which usually increase as time elapses. He will normally prefer to make his planning application in outline initially, leaving details of the development to be submitted and dealt with as 'reserved matters' at a later stage, after outline consent has been obtained, rather than submit a full planning application at the outset. In practice, however, the local authority will frequently and legitimately need more than a simple 'red line around the site' type of plan (which is basically what an 'outline' application consists of) in order to deal with the inherent proposals adequately and sensibly.

For a development scheme of any substance, the defined site area (including land not under his control or ownership shown edged blue) plus the written description on the application form will need to be accompanied by drawings showing the general form and extent of the building(s) proposed, in both plan and elevation, together with means of servicing, access and parking provision, plus an indication of fenestration, materials, skyline and landscaping. If the development site is within a conservation area, and/or affects a listed building, full application details will be required, in the form of a listed building consent (LBC) application. For a development scheme of major content and/or visual implications, or affecting a conservation area or listed buildings, or in any particularly sensitive or prominent situation (such as an urban fringe or rural landscape of recognised visual beauty or sensitivity), still more visual information will be required, usually in the form of perspective views ('artist's impression') and/or models and not just the small-scale 'block' variety of model).

Thus the design process starts at a very early stage in terms of submission of information to the Chief Planning Officer/Department. The 'design consideration and advice' process, to whatever degree it is applicable, should also start at much the same time, and a good local authority should provide adequately for this in terms of senior officer-level discussions.

But developers and their agents – architects, surveyors, etc. – may need planning and design guidance preferably prior to producing even informal proposals for discussion, and this of course is the role of the planning or development brief, followed by design guidelines (the principles of which we explored in the section on development plans, p. 34).

The precedent has been established, via a

legal case involving the London Borough of Richmond, that local authorities are in fact entitled to levy an extra fee charge upon applicants, additional to the standard planning fees prescribed by the DoE, for advisory discussions taking place before the submission of a formal application. Normally these extra fees apply only to developers and their 'commercial' schemes rather than to local residents and their 'householder' applications – a blanket requirement would surely be politically unacceptable, even though extra officer time may be involved in both types of proposals! It is quite often the 'commercial' proposals which require the most attention however, so the extra fee basis is sound.

As stated earlier, not every site will warrant either a planning or development brief – this will depend on the planning complexity and policy provisions for the site, as well as the magnitude or implications of the proposals. But some form of design guidance should surely be appropriate for any development site of substance, most particularly where its context is a sensitive one.

The format of such guidance, if adopted in principle across all or most authorities, could be that of a planning/design checklist: i.e. a list of policy and design considerations, already laid down in approved form by the local authority and identified on the list by a system of reference numbers. Clearly, it is of great importance that developers and their architects and other professional agents be provided, at a very early stage, with an informative indication of the Planning Authority's expectations for the area or site concerned.

The DoE, together with the RIBA and RTPI, was jointly considering the practicability of design guidelines in December 1990, in the form of model guidelines initially, formulated by a few selected local authorities. The Prince of Wales also strongly endorsed the idea publicly, at an RIBA Planning Awards ceremony in February 1991. The concept has also been explored by Judy Hillman, for the Royal Fine Art Commission, in her excellent 1990 booklet *Planning for Beauty*, in which she too encourages local authorities to draw up local design guidelines to use as a checklist, without dismissing the 'beautifully eccentric or brilliantly unexpected' (to quote the then Secretary of State for the Environment), and to give local people a role in the decisions about their surroundings. But here we come to the nub of the matter.

If local people – either individually or represented by their local authority – are to have a meaningful say in the decisions on development proposals in their area, then *this also implies some form of assessment (involving inevitably a degree of subjective judgement) of those proposals, additional to the 'design and/or planning checklist' set out by the authority*. And here we are back full circle to the traditional vexed debate about aesthetics versus control. The debate is indeed inevitable, and no amount of talk about 'architects freedom' will alter that; sooner or later, people (including council members or planning officers) will feel the need to make subjective comments or judgements upon the 'design merits' of certain proposals, above and beyond the design guidelines provisions; and this will have to be addressed as part of the process. Whether this process leads to progressive or regressive design solutions will be dependent on many (mostly localised) factors.

One extra piece of planning machinery exists, however, which has been found valuable in dealing with applications in sensitive areas and/or of debatable content or form, and that is the 'environmental statement' (also called 'environmental impact study'). This consists of a requirement to provide good reasoned evidence of their acceptability in terms of local 'impact' physically – not only as written argument (too capable of empty or ambivalent semantics on its own) but also in the form of supporting sketch plans, showing viewpoints, sightlines, proximities, height differentials, etc. This in the end is helpful to all sides – applicant, local authority and objectors – as it promotes a clear, informed and analytical approach to the three-dimensional building proposals with reference to their context (and assists in arriving at a clear assessment and decision in each case).

All proposals should of course show clearly their relationship to their surroundings: in many cases this is of equal importance as the proposals themselves, and in turn should also contribute to the design concept. But in sensitive areas, which can be primary areas such as town centres, or conservation areas (in the case of many of our historic towns, both apply) extra information and measures such as these 'environmental statements' are particularly relevant. And in the case of design schemes that run the risk of being regarded as unusual or controversial, the designer thereby gets an opportunity to explain to the public and the local authority the thinking and principles underlying his proposals. Many architects (and other building designers) could in fact help themselves more than they do, by providing some form of explanation or justification of their proposals, in the context of the site locality. If they fail to do so, they cannot complain if problems ensue.

There should be mentioned at this point another major system of local authority control – that of Building Control. This system runs partly parallel and partly in tandem with the planning system, and administers the Building Regulations, via a near-comprehensive set of detailed controls over constructional and public health and safety standards. It is not intended to examine this system in any depth here, as it is largely technological in content, but it is worth noting some particular aspects. It is entirely separate and different from the Town Planning system in content, powers and procedures, though some applicants mistakenly believe that a planning consent confers a Building Regulations approval – this is not the case. The latter must be obtained for all new or altered construction (except as regards a limited range of 'exempt' structures of a minor nature). Basically, construction details have to comply with the Building Regulations (the system is not open to appeal) though some flexibility exists as to alternative possible methods. Breaches of control can, as with planning, result in enforcement action, but with court cases under the Public Health Acts in respect of Building Control, where relevant. *A major procedural difference between the two systems lies in the 'deemed approval' provision,* where no decision has been issued under the Building Regulations within 5 weeks (or 8 weeks if extended), *in contrast to the 'deemed refusal' provision in planning* for failure to determine the planning application within 8 weeks.

In terms of design, the constructional requirements of the Building Regulations can have varying effects upon the form or appearance of buildings, though as we have said some degree of flexibility does exist with regard to alternative techniques. The major requirements of the Building Regulations generally relate to structural safety, means of escape in case of fire, and other public health and safety provisions, as applicable.

Consultations are linked together with negotiations as a necessary part of the process of assessment of planning and design proposals. There are both statutory and non-statutory practices: under the county/district structure the Highways Authority, for example, must be consulted about any development proposals affecting a primary road (or sited within a 60 metre distance of such a road, whether directly affecting it or not), and the County Planning Officer must also be consulted in respect of any proposals raising strategic planning policy aspects, such as the provision of a major new shopping complex on the periphery of a town, or a large new office development in the town centre. Where there are conflicts with structure plan policies, the county would normally raise formal objection, and expect this to be taken into account and reflected in the determination of the application (or appeal). The County Planning Officer might also wish to comment on certain design aspects of the scheme, but it would be the role of the district council, normally, as the 'determining authority' to pass final judgement upon the application. There are also other statutory consultation obligations in respect of demolition of (part or all) of a listed building – basically with English Heritage plus the 'six national bodies' including the SPAB (Society for the Protection of Ancient Buildings), the Georgian Group and the Victorian Society.

For the more compelling and/or contentious design schemes in terms of either urban context or design propositions, or both, the Royal Fine Art Commission has a (sometimes vital) long-stop role to play. The RFAC is apparently mandated to 'call in' any such proposals that come to its attention (or are directly referred to it), in order to adjudicate in a semi-official capacity and to guide both the local planning authority and the applicants on aesthetic matters. Their influence is, from personal experience, generally beneficial. At formal commission hearings on planning proposals, council representatives who have thereby ventured into the deep waters of 'design aesthetics' but who are not well versed in this field are however at a disadvantage.

Amenity societies, residents' associations and local political parties are the 'other side of the coin', where the consultation process is concerned. Of these the amenity societies normally have the most specific 'design comments' to make, with the other two types of semi-official consultees (plus of course private individuals) being more concerned to lobby the authority on a wider range of more general considerations – these can range from 'neighbour development squabbles' (mainly house extensions), to the latest major office application in the town centre.

Local Objections/Representations

Where consultation with an aroused and aggrieved local public is concerned, we are in volcano country – sites and planning issues that have either been dormant or gently simmering for many years can erupt suddenly, and with not inconsiderable force, to the peril of all embroiled in the situation. Yet it would be a basic mistake simply to consider that assessment should always equal the sum of all the consultations. The Planning Officer has a very important individual role to play in respect of the assessment he or she makes, and thus the recommendation put to the committee. It may sometimes be felt when faced with solid and overbearing opposition from the local objection lobby that the best plan is to kick for touch and advocate refusal. *But the Planning Officer has an obligation both to him- or herself and to the local authority to give a personal and professional assessment* independent of other players on the field, or of the crowd. The local people may often be right, where a simple or straightforward issue is involved, but they do not always see the finer points of the game. The Planning Officer must recommend upon a balance of opposing factors, including questions of design and 'amenity'.

This raises an interesting point – the word 'amenity' (which we have thus far avoided) crops up in innumerable committee reports and decisions, but it is nowhere fully or properly defined in planning legislation – most probably because it cannot be satisfactorily defined. It is an even more amorphous concept than that of 'good design' (which this work does attempt to explain, or at least highlight, by way of subsequent examples and their analysis via certain principles). Yet the wording 'in the interests of amenity' is quite probably the most overworked phrase in modern English! Planning Officers and committees are particularly fond of it. We shall not attempt a definition here, but in a general sense it does have linkages with good design of the environment –

in that good environment does not just simply happen, but is usually designed to some extent or other. However, these are semantics.

Committee Presentation/Information

The attitude of planning committees towards the design aspects of proposals can range from rare enthusiasm through mild disinterest to that of outright opposition. Much will depend on:-

- the context;
- the content (including any form of 'planning gain');
- the form of the proposals; and essentially
- the degree of understanding of the scheme, in the widest sense.

In all this, the quality of presentation to the committee is of considerable importance. This again comprises several aspects, namely the visual/graphic quality and clarity of the applicants' submission drawings (and in the case of major or 'sensitive' schemes, models); the resources of the authority to provide the best site and design information display methods, such as video, overhead or slide projection, explanatory diagrams or analytical overlays in addition to the normal procedure of pinning up application drawings on display boards; and last but not least, the Planning Officer's skills at verbal and oral presentation and, yes, advocacy, for he or she is in a sense acting as the applicant's intermediary at the committee meeting, at the same time as being the council's professional adviser.

The principal instrument with which to advise the committee is of course the Officer's written report, but this is supplemented quite often by the Planning Officer (or the Chairperson in some cases) giving spoken guidance or commentary. Applicants or objectors sitting in the public gallery sometimes get rather bemused (or aggrieved) if there is no oral presentation, i.e. in straightforward cases where the written report before the committee says it all! So a few minutes spent in a simple summary of the main aspects and issues of each committee application is amply worthwhile, in terms of 'public relations'.

A few planning committees allow, and a few even encourage, developer's architects and their other agents on limited occasions to attend and explain to the members the content, nature and details of their proposals, including design aspects. This can indeed be helpful in the case of major or 'sensitive' schemes, or in a 'competition' situation, but it is an arrangement which cannot be extended to all planning applicants – otherwise committee meetings would last all week! (Some councils have instituted the practice of independent architects' advisory panels to help inform the process.)

The essential point is to give the members as much information as they need, in order to come to a decision on the proposals before them, and to supply or display that information (including site plans or photos) in as clear and illuminating a way as possible. Schemes should always be shown in their immediate (and often wider) context, via plans and elevations, to enable fully meaningful decisions to be made. Moreover, councils have the right to insist upon sufficient 'site context' information being supplied. Drawings, etc. should show much more than they habitually do, concerning relationships between new buildings and existing development, in terms of plan layouts, street elevations and spaces, and hard and soft landscaping. It is surprising that legislation does not require this. Planning permissions are worth a great deal of money to the applicants; investment in better information and presentation is time and money well spent, even if the application is refused and has to proceed to appeal.

Appeals

The DoE's Appeals system in this country, like the local authorities' development control system, is almost permanently overloaded and there appears to be little that can be done to relieve it to any significant extent. It seems an integral part of the British way of life – but the cost of democracy is high, in respect of the time implied. Almost anything can, and does, form the subject of planning appeals in Britain (compared with countries as diverse as Germany, Denmark and Australia, where clearer systems of land-use zoning – long abandoned here – make planning decisions much more straightforward).[23]

The implications for the 'design aspects' of appeal hearings and decisions can vary greatly, according to the nature, content and urban context of the proposals in question, and the prominence or otherwise of design in either planning or architectural terms (usually both). Normally Planning Inspectors are not overkeen to become too involved in pure design issues or in the closely related aspects of 'aesthetics', and they will often tend to be unsympathetic to local authorities who have opposed a set of proposals purely on 'design grounds', and thus held up bona fide economic development. Inspectors are conversely also unsympathetic to councils that acknowledge the design merits of a scheme under appeal, but oppose it on grounds of local plan or other policies, if the proposals can be shown as 'not having an unduly adverse effect upon interests of acknowledged importance' – such as perhaps via the provision of a high quality of design in development (which surely should be a generally accepted and desired aim in any event). Councils have a difficult furrow to plough at appeal when it comes to design and aesthetic issues. It is clear that Planning Inspectors still do not often rate such issues highly, compared with policy and economic arguments.

Legal Agreements/'Planning Gain'

Refusals and subsequent appeals can sometimes validly be avoided by legal agreements between councils and developers to secure extra benefits or advantages for the local environment through planning approvals. The commonest form of this is the 'planning gain' contribution, but councils must beware of imposing such requirements where they are not essentially relevant. Any 'planning gain' should confer benefits on the immediate environment or locality of the development scheme, and be positively needed in the context of the proposals. *Good design should not be seen as a form of 'planning gain': it should always be viewed a basic requirement, and not as some kind of idiosyncratic 'extra' imposition by the local authority or as a generous concession by the applicant.*

FURTHER MEASURES FOR GUIDANCE AND CONTROL

Public Consultation and Participation

These dual modern processes merit our attention as relevant, indeed central, to urban quality, but we need to distinguish between them. Consultation on the whole implies presenting predetermined options to the public for their comments and preferences to be heard. Participation involves people in formulating ideas and plans, as part of a creative process of planning and/or design.

When the Skeffington Report on *Public Participation in Planning* was produced in the early 1970s, most local authorities felt that *they* would be the ones who would be initiating the planning or in some cases the development of the relevant parts of their area, and who would in turn be consulting the people living or working in those areas, as well as negotiating with the developers involved. Twenty years on, however, rather different circumstances apply. Whilst local authorities still have a 'forward planning' role, in terms of policy formulation via the development plan system and local plans, they generally no longer have the same degree of direct involvement in development schemes or their promotion, in that they do not have by any means the same freedom of access to capital funding for development purposes. (Central government restrictions and policy changes since 1979/80 have increasingly seen to that, and have also forced many local authorities to disgorge much of the land holdings they had.) But, as bodies of elected representatives if not as 'environmental trustees', it should be borne in mind that local authorities have statutory duties to perform in respect of democratic consultation of the public, extending over a wide planning spectrum.

Unfortunately, local councils have fallen as a whole into a generally 'reactive' stance, rather than a 'proactive' approach, towards most individual development proposals (as against local plan policy formulation, which is often more positive), and the pattern of public consultation has followed suit. Indeed, local authorities generally follow a semi-formalistic path in this, whereby:

1 Applications are received by the council's planning department (after any initial developer consultation with officers).
2 The officers consult those persons directly affected, informing local/ward councillors at the same time; local amenity societies may also be consulted.[24]
3 Discussions are held if necessary between officers, ward members, and residents/others directly affected, then subsequently between the officers and the developer and his agents, to achieve any changes or compromises necessary.
4 The application proposals are then reported to the planning committee, with a suitable recommendation, for member decision (chairs of committees will have an increasing role in decision-making).

There are, however, limitations upon any local authority's scope for consultation on development proposals, and thus upon the public's scope for participation. In turn, this may be held also to affect the 'quality of the product', as the proposals will normally have been formulated by the developer on largely financial grounds, and will then have been subjected to a (sometimes lengthy) process of 'mauling about' at officer and committee levels before final decisions are made. Although the 'bumps' will have been ironed out in the end, some of the freshness and spontaneity of the initial design approach may have suffered, or been lost, in the process. It may also be questionable in some cases as to whether the proposals will truly reflect the needs, interests and aspirations of the local populace, i.e. those most directly affected.[25] Also the developer and his architect will have been subjected frequently to what they see as irritating and costly delays, while the ball is bounced backwards and forward between applicants, objectors, local authority officers and members, and back to the applicants again for reconsideration – and that is without mentioning the possible further stage of going to appeal, with commensurate extra delays and costs.

It is hardly surprising, therefore, that the alternative route of 'Community Architecture' was conceived during the 1980s. This was originally a concept geared mainly to reversing inner-city decay, whereby the developer and/or his architect would 'do the job of the local authority' by getting in amongst the local populace (whether they were directly or indirectly affected by the scope of the development envisaged) at the outset, not only to seek their views and canvass their support, but also to encourage their participation in the design process. This was the 'proactive' rather than the 'reactive' approach in action[26]. There was nothing very new in the developer consulting the local residents/shopkeepers and canvassing their support before putting his planning application in, though the practice became more evident in the 1980s. However, the scenario acquired a new turn when, following the lead given by

Prince Charles in spearheading this worthy cause, architects (of whom the major exponent of this particular approach was Rod Hackney, a consultee of the Prince of Wales) actively encouraged residents to become involved almost at the start of layout or building design formulation with ideas and schemes for the better planning and development of their locality. This 'hands on' approach to development is of course in sharp contrast to the usual local authority 'planning control' practice of consulting the public at the formal application stage. (It should be mentioned that long before 'Community Architecture' became fashionable, the Byker Wall residential scheme at Newcastle-on-Tyne had paved the way, by initially consulting the local community).

There is basically nothing wrong with the democratic principle of this approach, and certainly one should not be dismissive of it – many a residents' association has come up with thoughts, ideas and firm proposals which have given the local authority pause for thought and a fresh perspective. Nevertheless, the other side of the coin is that this process depends not only on the guidance of the developer and his professional agents, but also on the simultaneous involvement of the local authority's officers (acting in their capacity as the council's in-house professional advisers). It also lies within the framework of the decision-making process of the council generally – upon which rests the essential question of statutory planning and other consents. Otherwise the council will in effect be presented with a *fait accompli* by a financially motivated developer.

The local authority might well argue that its role is the central role, and that it is being bypassed; whilst the architect or the developer will equally well contend that the process of Community Architecture is more directly related to local needs and interests as well as marketplace economics and produces greater initiative and more 'true' results in the end. Ironically, the process has come under a certain amount of disapproval from some of the more purist factions of the architectural profession itself, in terms of their tenure of the 'high ground' of aesthetic conceptualising, i.e. as opposed to the down-to-earth, bricks-and-mortar results of the Community Architecture philosophy of 'giving the people what they want, not what we think is good for them'. The words in quotes are of course those of Prince Charles, to whom the concept of Community Architecture (like his espousal of Classicism within the Great Debate on 'style and taste') was firmly linked. We should accordingly note this concept and its philosophy, before passing on to other considerations, but not before making the comment that the relevance and benefits of genuine and constructive public consultation, in whatever form, must be apparent to all, as far as urban quality is concerned. None of us can afford 'castles in the air' these days!

Sensitive Areas/Complex or Major Schemes

It is self-evident that, when faced with a particularly complex/major scheme, or when dealing with substantial proposals in a sensitive siting (e.g. a town centre or residential conservation area, or a prominent location open to a wide range of views, or affecting a particular historic building), it may well be necessary for the local authority to call in expert advisers (whether for commercial, architectural, landscape or antiquarian advice).

Adequate budgetary provision should (in theory) always be made for such eventualities, in the form of 'contingency funds for specialist services'; though forecasting exactly how many compelling needs or projects are likely to arise in any financial year for such attention probably belongs to the realms of clairvoyance.

The larger redevelopment schemes will of course attract larger planning fees, which in turn can and should be used (at least in part) towards employing specialist consultants to advise the authority. Probably the most frequent and yet most unpredictable type of case for specialist treatment concerns historic buildings (or areas), where the wrong approach can spell disaster for the building or area in question. Historic buildings are particularly apt to create sudden and unforeseen problems (usually related to original deficient structure), which then demand skilled and often rapid remedial attention.

Contingency provisions within local authority budgets will continue to play an increasingly important, though unquantifiable role in protecting our environment, whilst competing with more prosaic but predictable needs for available finance.

Developers, Architects and the Public

The whole question of taste and style in today's architectural and planning scene is increasingly fraught with difficulty, conflict and discontent. The reasons are not hard to find. First, there is the basic sharp divide between the traditionalist and the progressive schools of thought.[27]

Second, and more significantly, there is the marked trend towards 'pluralism' of view, both within the ranks of the architectural profession and the general public, i.e. the consumer society. The latter aspect is potentially the most telling aspect of all. Increasingly, people have been encouraged to believe in the twin concepts of 'money talks' and 'the right of free choice' in almost everything. At this point, it is worth quoting the architect Charles Jencks:

> If the claim for more choice increases further, as is likely, community groups will soon be asking for control over their local environment, and the natural consequence of this should be the right to vote on contending architectural schemes. At present, democracy tends to be limited to a vote by professional [sic] representatives who are appointed. Characteristically, a developer will hold a closed competition and choose a winner, as in the Paternoster scheme; but there is no reason why the public can't be brought into this process and make choices at several points. This will increase the problems for the architects and developers at the beginning and it will increase dissonance; but it will also lessen the troubles at the close of design, decrease the number of appeals, and produce an architecture that is closer to what the locality wants. Just as importantly it will shift the architect's attention from the planners' and developers' tasks to those of the ultimate clients, the users. At least in theory it should.
>
> (Jencks 1988: 30)

Although the days of local authority development are largely over, and hence the need for their own competition processes to select developers and their architects is rather more limited than in the 1960s and 1970s, there is still the need for some form of selective process to ensure a sufficiently high standard of development and quality of architecture. That process clearly needs to take place initially within the private sector, but with proper provision for public involvement and consultation if difficulties are to be avoided or minimised when the development proposals arrive fully within the public sector (i.e. the local planning authority) for assessment and decision.

It is the prerogative of the developer to choose his own architect and he will hopefully do this on 'track record' (experience and competence rather than on questions of 'style' and 'taste'), whilst having more than half an eye on the question of how well that architect's work is likely to be received by the planning authority and the public. The onus is upon the applicant/developer to ensure that the right professional adviser(s) is/are engaged for the job in hand.[28]

Some modern architects have been brave enough publicly to dissent from 'design intervention' by local authorities, and the watering-down process by planner that goes with it, on the basis that they see design as an integral professional matter, very largely personal to the architect himself or herself. It is surely wrong that such architects (as Michael Manser has from personal experience pointed out) should suffer loss of work due to ostracism by Planning Authorities offended by their resultant outspokenness, however much one might deplore an arrogance of approach. *It is also untenable in today's 'pluralistic' society to expect architects to toe a single preconceived line when it comes to the vexed, and essentially misleading, questions of 'style' and 'taste' in architecture.*

Planning and Design Guidance Material

Many district councils have produced a range of guidance leaflets on subjects as diverse as trees, historic building grants, domestic extensions and shopfronts, mainly for the benefit of their residents and the business community locally. Some county councils (e.g. Surrey County Council) have also produced guidance leaflets on conservation/historic buildings, and some on landscape and rural conservation. These are only examples.

Although the DoE discourages local authorities from becoming involved in detailed design aspects of planning application schemes, most councils rightly concern themselves with the close control and 'management' of detailed aspects of the urban fabric (particularly in historic areas), in the interests of preservation, quality control and good maintenance of the environment, and the further encouragement of the same via grant aid, where applicable.

One might well ask in passing why, if it is legitimate for local authorities to concern themselves closely with the detailed treatment of the existing urban fabric and environment (in the interests of urban quality generally), it should not also be legitimate for them to take a close interest in the detailed design aspects of new proposed development. Anyone who knows anything about design quality (urban or otherwise) knows that high quality of design depends just as much on sensitive and appropriate detailing as on the overall concept or 'massing', important as the latter may be. Whilst a good, responsible and competent firm of designers will produce sensitive, well-detailed schemes, there will be instances where this is not the case, and where the detailed intervention of the local authority planning design staff will be merited.

One planning authority responsible for a commendably wide range of advisory leaflets for its area has been Hove Borough Council, in Sussex.[29] Hove is largely densely developed and historic in character and, like its neighbour Brighton, is tightly contained between the sea and the South Downs. The whole of the Brighton and Hove area (now to be merged as one local authority) is subject to continuing pressures for change to the urban fabric, since expansion is severely limited by external barriers and internal constraints. To assist in dealing with these pressures (which if left unguided would quickly degrade the urban environment, as uncontrolled market forces nearly always do),[30] the then Hove Council prepared in the early 1990s an impressive array of guidance leaflets, highlighting relevant powers of control and their applicability.

In residential areas, the most common cases for treatment via local advice are extensions, conversions and alterations. Extensions to the rear of properties usually raise questions of overlooking (privacy) and infringement of light angles (daylighting) and are the most usual problems, controllable by setting down minimum distances from side boundaries and 45° angles of light restriction. (Hove Borough Council's policy stipulated that conversions should normally be carried out within the existing shell of the building but that where an extension is proposed, the Council's booklet *Extensions to the Rear of Properties*, embodying the above considerations, should be consulted.)

In respect of conversions and alterations, Hove Council decided to produce a comprehensive guide as well as a series of leaflets, since the borough's numerous squares, terraces and residential streets contain much 'flatted' accommodation, which is subject frequently to both internal and external change. The guide plus leaflets together cover many detailed physical aspects requiring firm and sensitive control, such as:-

- doors and windows, including windows in roofs, e.g. dormer windows (with special reference to detailed design);
- roof repairs (with particular reference to materials);
- painting facades and textured coatings, with a specified colour range for painted stucco in Central Hove (textured masonry paint is not allowed on historic buildings);
- aerials, cables, flues and pipes (i.e. 'modern building paraphernalia');
- architectural details (e.g. balustrades, cornices, balconies and chimney stacks; their repair and maintenance);
- fire escapes (with particular attention to the position, design, and colour – normally black gloss – of external fire escapes where these are necessary);
- front gardens (treatment of curtilage walls, and restriction against use for car parking);
- refuse disposal (suitable forms of bin storage);
- car parking provision (off-street/on-street).

Powers of control are of course much stronger in the case of listed buildings, conservation areas, and Article 4 Direction areas, all of these being statutory based.

(Note: Article 4 Direction orders remove 'pd' or 'permitted development' rights in designated areas, and enable similar degrees of control over potentially disfiguring urban characteristics, as in conservation areas. Thus in some Article 4 Areas, certain alterations to buildings which do not normally need planning consent – e.g. painting facades – now need consent (as in areas of Hove such as Adelaide Crescent, Palmeira Square and Lansdowne Square). Another Article 4 Direction exerts control over certain additions to facades – e.g. steps, hard-standings, porches, replacement roofs and windows, garden walls, etc. – in other parts of Hove (namely large parts of Cliftonville, The Drive, and Denmark Villas conservation areas).)

In addition, Hove has defined certain Areas of High Townscape Merit (as 'conservation areas of the future'), in which planning applications are carefully examined to ensure that new development or alterations do not dilute the architectural character of the area. Leaflets describing these areas, with design advice, are available from the planning department, as is also the explanatory leaflet on Article 4 Direction Areas. These are examples of good local authority practice.

The 'flagship' of statutorily based protection of historic environmental areas, however, remains the designated conservation area, together with the listed building, and all good councils will normally have leaflets explaining the relevant powers pertaining to these. As already outlined in a previous section of this work, the demolition or partial demolition of a property in a conservation area requires planning consent, while some minor alterations or development which would not need consent elsewhere will however require it in a conservation area. Demolition of a building, or removal of part of it without consent, is an offence. Trees in conservation areas cannot be lopped or felled without consent. Materials are strictly controlled. Satellite TV dishes on listed buildings and in conservation areas are likely to need consent, and so on. In addition, a primary function of such leaflets will be to define the local authority's designated conservation areas on Ordnance Survey Map extracts; a function also provided by the authority's planning handbook. *All these types of documents are indispensable parts of a Planning Authority's repertoire for the control of the environment and the maintenance of high urban quality.*

Hove Borough Council also produced an excellent booklet on *Shopfronts in conservation areas* (September 1990), again like its other design control publications fully illustrated with well-drawn sketches and diagrams, giving detailed guidance from overall shopfront elevations down to individual architectural and other details, backed up by relevant legislation and detailed local policy.

As regards planning handbooks, their 'mainstream' functions are to give a general guide to the overall characteristics of the authority's area; to explain the structure of the planning department; and to give detailed guidance on the planning system operationally (in terms of structure and local plans, development control, building control and conservation, together with grants and fees).

Planning handbooks are also an invaluable feature of a local authority's activities and services: together with advisory leaflets of all kinds (extending also to services such as housing and environmental health) they provide a meritorious example of 'public sector responsibility and enterprise'. In these days when it is politically fashionable and expedient in some quarters to minimise the role and value of local government and the public sector generally, it is as well to stop and consider the wide range of positive contributions provided to the public, such as those outlined above.

We pass now to other measures which can have an equally influential, though broader, effect on the quality and shaping of the environment from within the planning system.

Land Use/Floorspace

The old development plan (town map) system set out in annotated map form clearly defined sub-areas of differing land use within each local authority's area (viz. residential, industrial, educational, business (offices and commercial), civic uses, public open space, etc.) These sub-areas either reflected the predominant existing uses or indicated the proposed primary uses of each area or 'parcel'. There were, of course, certain broad areas of principally residential or open space uses, together with areas known as 'white land' (which were expected to remain largely undeveloped, or were held in reserve in case other development land ran out). Areas of open land were also designated for extra or special protection from development, such as AGLVs (Areas of Great Landscape Value), or Green Belt (land ordained to remain undeveloped and largely open, with the principal intention of preventing the uncontrolled spread, and in some cases coalescence, of urban settlements, and of maintaining a clear identity and distinction between town and countryside).

These mechanisms have been very successful in regulating and controlling the extent, relationships and to some degree the character of the built-up and open areas. In particular, the Green Belt has been especially effective in holding in check the increasingly strong tendencies and pressures for outward spread of urban areas and random 'colonisation' of the countryside.

The major mechanisms have in effect been transferred without dilution from the old town map system to the local plan system (now the unitary plan system), though the previous pattern of clear-cut and therefore fairly strict 'land-use zoning' of the town maps has been superseded under the local plan system by a shift towards written policies and away from 'zoning'. An exception to this, to some degree, has been the continued clear definition of 'Primary Office Areas' (POAs) within local plans, though the government's relaxation of the 'B1' business category of land use to allow greater flexibility (i.e. permissible change of use) between the different types of commercial and office use has diluted the planning authority's control over the growth and spread of office use, in terms of both quantum and location.

This, together with the extra overall pressure upon central urban and residential areas caused by the tight maintenance of the Green Belt, means that we now have an urban environment increasingly under siege, in the form of residential densities remorselessly increasing due to 'infilling' and domestic extensions, and of Central Areas becoming more and more prey to unwanted or excessive amounts of office and other commercial development. The recession of the early 1990s may give some temporary relief from development pressures, but in the absence in Britain of proper regional control of planning and economic growth, the problem will not disappear. Some judicious relaxation of Green Belts is now taking place here and there, but understandably this is mainly limited to special or overriding cases, such as redundant hospital lands (the Epsom Hospitals Cluster in Surrey is an example). Accordingly, 'quality of life' can all too easily suffer from the effects of overdevelopment or undue intensification, in the form of generation of excessive traffic activity, parking and roadscape demands, for example.

Urban quality may, in a strictly functional or hygienic sense, gain from the application of modern standards of space or accommodation; but 'urban character', which is even more germane to this thesis, usually owes little to the imposition of modern standards – it is more endemic. We have all seen old, perhaps tired, but still characterful areas or streets, with a continuing lifespan, threatened or overwhelmed by injudiciously located modern development of an alien use and character, and a totally alien scale (e.g. 'the seven devils of vast, impersonal unresponsive concrete', as A.R. Woolley wrote of the redevelopment of the Little Gate/St Ebbe's area of Oxford, in the *Clarendon Guide* to that city).

Even if out-of-scale or incorrectly located new development can nowadays be more seductively dressed up in pseudo-historic or high-technology facades (it is a matter for debate as to which is worse), the problem of incompatibility is still essentially the same. The twin underlying culprits are of course land values and market forces: unless firmly guided in all cases by perceptive and responsible local land-use planning, the effect on the environment can be unfortunate to say the least. Some flexibility, however, can legitimately exist in the case of small, 'in-scale', good quality, non-conforming uses which can be shown to create no harm to their surroundings. But the extra possibility of 'planning gain', which is frequently attendant nowadays on substantial new development proposals, can be either an actual benefit or a false friend, as we have discussed earlier.

One useful weapon that is still to hand in the control of floorspace quantum, is the Floor Space Index, or FSI. This is the area of the total floorspace of the buildings on any given site, divided by the area of the site, including half the area of any roads adjoining it. As Lewis Keeble said in his *Principles and Practice of Town and Country Planning* (Keeble 1952), the FSI survey can only usefully be applied to Central Areas of considerable size (or subsidiary centres in some very large towns), where the quantity, variety and congestion of non-residential uses are sufficiently great to necessitate an FSI survey prior to formulating plans for redevelopment.

The FSI method of controlling floorspace total was set down in great detail in the Ministry of Town and Country Planning's handbook *The Redevelopment of Central Areas*, first published in 1947. As a numerical tool, it had great applicability potential and also considerable implications for the closer control of the density and even the form of Central Areas. Questions such as the relationship between traffic generation and floorspace, and the growth factor in each, were encompassed. Recommendations were made as to desirable levels of FSI at that time for different types of land-use zone (e.g. 1.5 and 2.0 respectively for shopping and offices (business zone); and 2.0–2.5 for wholesale warehouses and light industrial uses respectively). FSI surveys, whilst time-consuming, would establish much more clearly not only the overall amount and intensity of use of floorspace for an area, but also how it was made up, taking into account the various different constituent use classes, whilst not distinguishing between minor sub-variations of use, multi-occupancy, or physical irregularities of building plan forms which often have some bearing upon the degree of over- or under-use of the floorspace involved. (Lewis Keeble set out, in his ever-practical way, a simplified method of carrying out an FSI survey without undue expenditure of time and money).

Although the use of Floor Space Indices can provide a fairly realistic guide as to the present amount and usage intensity of existing floorspace, and thus provide a basis for proposals as to future use, intensification or redevelopment of an area or block, it is a tool that is nevertheless seldom used these days. Apart from the time–cost labour implications for hard-pressed local planning authorities, its real 'fault' is perhaps that it is too specific and inflexible for the aspirations and accommodation of today's development forces – though in simplified form it could still have a role to play in local authority control of commercial development, where the urban context is at all sensitive.

High Building Strategies/Building Height Policies

A particularly intrusive element in the traditional urban scene is the high building. The intrusion can be for good or ill, depending on a number of things, but the numbers, location, function and even presence of high buildings should be a matter of prime interest and concern to planning authorities. To this extent, every authority faced with the incidence of such elements within its townscape should prepare, or already have prepared, a High Buildings Strategy against which all planning applications involving buildings or structures higher than the maximum average 'norm' for the areas in question should be judged. The High Buildings Strategy should be interlinked with a Building Heights Policy, which should form the basis for consideration of all building heights, or height ranges, within the urban area. It is surprising that more local authorities do not prepare such policies as a form of 'insurance'.

In townscape terms, there is hardly anything more significant for the character of an area than skyline, which is in itself a complex product of building height, form, function, style and type of roofing materials, though mainly it is a matter of building heights and their juxtaposition. It is the skyline which has given so many different towns their distinctive character. In turn, this is why well-placed, well-shaped high buildings can sometimes assist rather than detract from the townscape and skyline of an urban area (but also why a single bad mistake in this respect can spell disaster to the appearance of that area). Thus, high buildings can be used to reinforce image and presence, e.g. of a Central Area, or to emphasise topography, as in the Birmingham Urban Design Strategy (Stage 1) by Tibbalds Karski Monro Consultants.

A Building Heights Policy first will have regard to the predominant character of each locality, in terms of existing land use, building size, height, street pattern and area density and layout. From this, it will assess what is the comfortable/natural height range or maximum for that area or locality and formulate a policy accordingly. (The question of buildings above and beyond this height range or 'norm' will be treated separately, within the High Buildings Strategy.) The Building Heights Policy will have to be fairly simply expressed in order to be workable – but should not be oversimplified to the extent of eliminating any degree of flexibility in practice, and of thus creating a rigidly uniform and boring skyline, without variation or articulation. This is why an acceptable height range for each area is more successful in townscape terms than a single rigid limit. It may, for instance, be appropriate to suggest a 4–storey 'norm' for commercial uses in a tradi-

TOWN PLANNING FRAMEWORK 65

Figure 10 Lewes/Brighton: *rus* and *urbs*.
The views of Lewes, Castle Banks (left) and Brighton, Sussex Square (right) make an interesting comparison. They are excellent examples of *urbs in rure* and *rus in urbe* respectively and even have a historical and visual affinity in their formation and layout. The humble terraced dwellings arranged along the contours below Lewes Castle enjoy enviable views and the use of the green valley hollow; while the sophisticated Regency crescent of Sussex Square has a similar relationship with its own green urban space. It may be more than coincidental that Amon Wilds, the builder used by the developer for Brighton's Kemp Town, came from nearby Lewes.
Source: Michael Parfect

tional town centre, but its Building Heights Policy would be better expressed as a permissible 3–5 storey range (with 5 storeys as the exception) rather than as a uniform 4-storey maximum. Indeed, the largely historic high street frontage may well be predominantly 3-storey, in which case some new buildings within that frontage may need to be kept down to that height; whilst others may be appropriate as a mixture of 3 and 4, or even 4 and 5 storeys (depending on site sizes and street elevation).

A High Buildings Strategy should be the natural counterpart of the Building Heights Policy. As already pointed out, a high building or structure can either benefit or detract from a traditional skyline, depending on form, size and location principally. Church spires and towers sit naturally within virtually all types of townscape, from city centres to residential areas; but 'tower' housing and office blocks (particularly those of the 1960s) in general do not. Fortunately, the pendulum has now swung away from the construction of high-rise housing blocks in Britain for sociological reasons and some have already been demolished. Certainly there is a general confirmation of and return to low-rise living, though with higher densities than many people previously have been used to.

The particular impact of high buildings upon the urban skyline is such that their location must be both justifiable and appropriate, and also that their grouping, and their combined effect, must be carefully considered. A high structure or building immediately suggests (or should suggest!) an appropriately important location rather than a random placement. Groups of high buildings should thus be even more sympathetically and justifiably placed; their presence and effect *en masse* on the skyline is of unambiguous significance. The city centre of Cape Town, South Africa, provides an excellent example of confinement of nearly all its high-rise commercial buildings to within a clearly defined Central Business District. On the Continent, some historic cities, such as Prague, have over the years intensified their city centres without sacrificing their skyline to modern tower blocks. A view of Koblenz in an etching of 1632, for instance, shows a cluster of church spires, some in pairs, set among close-packed streets contained within a series of towers atop the old city walls – a delightful historic scene. St Paul's Cathedral in London also rose from a forest of spires in Canaletto's day – but is all but dwarfed and immersed now in an alien forest of modern high-rise office blocks. Market forces clearly were too strong, and planning was absent.

Similarly, on a more local scale, a cluster of Regency streets and terraces in the Montpelier district of Brighton descends gracefully towards the sea. One terrace focuses charmingly on a Victorian church spire. In the background, however, an 18-storey, lumpish, late 1960s/early 1970s tower block of municipal apartments rises heavily into view, an alien feature, immediately detracting from the scene. This is an example most probably of financial expediency (the flats were integral with the 1960s Churchill Square shopping centre, and no doubt formed part of the 'development package'), but it is also an example of visual insensitivity, in the massing and detailed handling of the housing block. Progress? Doubtless the tower block (Chartwell Court) made good economic sense at the time, but it does not make good townscape sense by any means. Churchill Square is now (1996) to be redeveloped as a shopping centre, but the residential tower blocks are more than likely to remain.

There are other high modern blocks which together dominate the Brighton seafront, and the still predominantly Regency and Victorian town. The original scheme for Churchill Square in fact proposed a second block of 18 storeys and third block of 30 storeys on the seafront. (Exceptionally, the large new American Express building of 1977 in Edward Street, Brighton by Gollins, Melvin & Ward succeeds by avoiding the vertical tower block formula.) These comparisons give only an indication of the damage that can be done to the environment by insensitive provision, location, scale and handling of high buildings, and of the value of their intelligent placement and grouping within the urban landscape, as well as of individual design.

The economic pressures for excessive floorspace amounts can result in lasting and often irreversible damage to the life and character of towns – not only in visual terms. These pressures must be controlled, counteracted and pre-empted as necessary by clear and firm policies not only for high buildings and building heights, but also for control of the floorspace quantum itself, whether in the form of high-rise or low-rise development. The damage can come in either form!

In examining these further measures for guidance and control of the urban environment, relative to the planning system in this country, we should not lose sight of the underlying (or perhaps overriding) need for local authorities to have an urban design concept for every relevant part of the areas they administer, i.e. where change can be expected of an order or type that is likely to affect those areas to a significant extent. This may be no more than the effects of development (full or partial) of an individual site within an area of recognised urban quality, which is important to maintain – or it may at the other extreme concern the redevelopment of whole areas where both the financial and the environmental stakes are high, and potentially in conflict.

The enlightened architect (and even developer) should also be thinking in terms of urban design concepts, though whether these will necessarily accord with the local authority's thoughts is quite another matter. All too often, as experience in one British city after another has shown from the 1960s onwards, the exigencies of 'market forces' have ruled the day, driven ever onwards by central government dogma, and by developers and accountants (probably most of all by the latter, since not all developers have displayed a totally monetarist approach devoid of urban design sensitivity).

TOWN PLANNING FRAMEWORK   67

Figure 11 Amex Building, Brighton: context and quality.
The American Express offices, though dating from the 1970s, are one of the best modern buildings in Brighton. Not only of Classical, pagoda-like elegance in mass and detail, the building enhances a wide area of urban background generally. It pulls together a fractured tableau of banal tower blocks, provides a visual focus for the area between the Steine and Kemp Town, and makes an interesting comparison with other major buildings of individual character such as the Dome and the Royal Pavilion. Unloved by preservationists, it is nevertheless an example of the need to consider the wider context, as well as the closer/more innovatory aspects of any building.
Source: Michael Parfect

The Prince of Wales, in his various pronouncements upon the patchwork nature and poor standard of development in this country in modern times (and, by implication, the failure of the planning system and both planners and architects to stem the tide), has placed his finger squarely upon this problem – without, however, recognising the importance of the public sector role *per se* in liaison/co-operation with the private sector. What else can explain the way in which the City of London has been overdeveloped commercially, to the point of extinction of its former pre-eminent historic Wren skyline, by a mass of high-rise office blocks? Two things are clear. Evidently, there was a process (presumably unplanned) of progressive intensification of the commercial office floorspace content of the City via a series of successive permissions for high-rise blocks, which came about owing to: -

- the lack of any policy originally for at least partial decentralisation of office growth into new external or peripheral areas, coupled with
- the lack of any commensurate policy or mechanism (such as a Building Heights Policy or Floor Space Index or plot ratio controls) which would have held down such office growth as was still deemed necessary and permissible within the historic City confines to manageable levels, in scale with the inherited environment.

It is self-evident that the explosive growth of market forces in the City took place in an uncontrolled fashion chiefly owing to overwhelming economic arguments but also owing to the lack of planning for, and provision of, suitable alternative site locations (such as we are now belatedly seeing at Docklands and Canary Wharf, east of the City). Up until 1990 or so, the city of Prague (due to 40 years of keeping capitalist forces at bay), had successfully preserved its historic skyline and the beauty of its historic environment generally by restricting high-rise commercial blocks to mainly peripheral locations where they would pose no threat to its integral character – whatever the shortcomings in other respects of its enforced State system. London, in complete contrast, by capitulating to the short-term exigencies of market forces and commercial pressures, has succeeded in virtually completing the process begun in the 1950s and 1960s of engulfing in modern tower blocks the historic townscape built up by Wren and others from 1660 onwards, when the great dome of St Paul's and a rich forest of church spires (of both English and Dutch Classical/baroque form) dominated the skyline, but nevertheless remained in touch with human scale.

The process of overdevelopment of the City of London no doubt started in Victorian times, when plot ratios[31] of 5:1 or 6:1 supplanted the more 'humanistic' traditional figure of approximately 3:1, resulting in narrower, more tightly built-up streets than hitherto existed. It was in the 19th century, too, that the 'ideal' building heights of 4–5 storeys for most urbanised situations (already discussed) began to be exceeded, adding further to the 'canyon' effect of City streets – to be made more horrendous by the explosive 20th-century growth in traffic. (The City of London has now mitigated the situation by directing much of its 'through' traffic away from the denser, tighter network of historic streets in the City 'core' area.)

Yet in terms of urban skyline, the real damage was not done until comparatively recent times, when the legacy of the late 17th century and the 18th century (appropriately called 'The Age of Enlightenment') was submerged by the forces of commercial gain, and when no alternative answer apparently was available. Had decentralisation been planned for, and was sufficiently feasible, and had sufficient local controls such as plot ratio and building heights policies been brought adequately into play, the picture might have been very different as far as the City was concerned. That unique skyline (again we stress it) of church spires and towers 'reaching up into the ever-changing sky from a rich matrix of gables, cornices, roofs, chimneys, turrets and pinnacles' (in a way which modern development can rarely if ever match) is now lost in terms of the wide external views, even if still available in limited localised views.

Whilst the next major tranche of burgeoning modern office growth now amasses relentlessly east of the City, in the heart of Docklands, the City itself is left with its one last chance (as Prince Charles put it) of redemption, via the redevelopment of the Paternoster area north of St Paul's. The auguries initially were not good. With the exception of one early competition scheme (that of John Simpson), the requirements inherent in historic urban character and context had again been all but fully subjugated to the overblown demands for excessive amounts of office floorspace, so that whatever way the competing architectural practices tried to arrange the total accommodation on the site, a maximum amount of monotonous floor space and an excruciatingly small quantity of public open space resulted. Thus, as Leon Krier commented,

> All this [design] work had indeed little to do with the proper meaning of master-planning; it is but slavish and subordinate to short-sighted pressures and greed. In all great cities, the plot ratio of 3:1 was hardly ever exceeded. This density is comfortably reached by buildings not exceeding five floors, allowing for private and public gardens, generous public spaces, well-lit and beautifully proportioned streets and squares.
> (Krier quoted in Jencks 1988: 51)

The danger is quite evidently that once the authorities allow utilitarian building speculation to break through the 3:1 plot ratio and the 4–5 floor height limits, the inevitable outcome is that land values literally go through the ceiling, bringing in turn pressure for ever higher plot ratios, thereby creating a vicious circle.

It is clear that, in order to maintain a human-scale urban environment, the ball is firmly in the court of the City planning authority, to achieve it by whatever means it legitimately can. The revised proposals for the Paternoster area, by a group of architect-planners, including Robert Adam and Quinlan Terry, working from the precedent set by John Simpson's initial 'Classicist' scheme (the only scheme which had received favourable comment from Prince Charles), gave greater hope that a total package of Classical-style building developments together would reflect (a) the original historic street pattern (to a large extent) and (b) an appropriate and humanistic scale and style of architecture, worthily to replace what was destroyed in the Blitz of 1940, and also to relate to the still supreme presence of St Paul's Cathedral, Wren's greatest single legacy to London.

Pressure to increase the total office floorspace from the Victorian level of 60,000 square metres to some 100,000 square metres has fortunately been resisted, and a scheme reverting to the former has been produced, resulting in plot ratio of only 4:1 rather than the grossly unacceptable figure of nearly 10:1 which had been threatened by previous proposals. The new scheme relates satisfactorily in scale to St Paul's (5 storeys maximum for buildings facing the Cathedral); the bulk of the office floorspace is contained in blocks of up to 9 storeys maximum, facing the wide sweep of Newgate Street; and a new civic piazza is formed centrally within the Paternoster area (replacing the long-vanished Newgate Market), led into by glazed arcades, loggias, lanes and streets of shops and cafes. This central square is planned as the focal point in a series of new traffic-free public spaces, lined with a variety of shops and services, restaurants, wine bars, etc.

NIMBY AND *LAISSEZ-FAIRE*

No examination of the problems of development control and urban quality control would however be complete without some reference to NIMBY and *laissez-faire*, two opposing spectres which consistently haunt the best professional efforts of planners in dealing with development.

Both the NIMBY (Not In My Backyard) syndrome and *laissez-faire* (an older phenomenon which still persists, in new forms) are closely connected and related to the modern concepts of economic determinism and 'market forces', the latter typifying the credo of the Thatcher years in particular: closely connected but contradictory. Whilst the free-market philosophy manifests itself in all manner of development opportunities and enterprises, it is ultimately planning control which must determine the amounts, forms and, indeed, the locations of development (leaving aside the questions of economic climate and money supply – both basic prerequisites). The element of contradiction that arises, however, is that those who most forcibly urge the promotion of market forces in terms of development are usually those who will fight tooth and nail to resist the intrusion of such development in their own personal environments.

Individually, the problems may not seem great in themselves; nationally, they amount to a time-absorbing catharsis in which the whole planning system, and society in general, is constantly employed. There seems no way out of the conflict between freebooting development pressures or private 'betterment' aspirations, on the one hand, and individual protective interests on the other. The planning system in Britain was set up largely to deal with such conflicts of interest, and is now well and truly saddled with all the practical implications; though were it not for the presence of a planning system of control and intervention (including quality control), the situation would have got out of hand to an unspeakable extent long ago. Two cheers, then (at least) for planning. Politically, however, the basic contradiction between 'NIMBY' and *laissez-faire* is a political conundrum (unsolved, and probably unsolvable) of the first water. The answer that will probably emerge, one suspects, is the traditional notion of encouraging profitable development (in the absence of directive powers via regional planning authorities) to locate itself within those areas that are seen to be both disadvantaged and in need of redevelopment and 'expansion'.

OTHER MEASURES FOR OBTAINING URBAN QUALITY

In a work of appraisal of urban quality in general, it is not the nature of this exercise to set out legislation in full detail or depth, but instead to refer selectively to it where it is particularly relevant to our analysis. The intention is chiefly to identify major principles (or inbuilt limitations) as to how far the planning system can facilitate enhanced urban quality.[32]

There are a number of other 'supporting' measures open to local authorities by which urban quality may be achieved, maintained or controlled, apart from positive environmental enhancement steps (which we shall look at in the section following), including the question of design guides, which lie part-way between being design control and enhancement measures.

These additional measures may be listed as:-

- traffic measures and parking provision;
- development phasing control;
- control of demolition/derelict sites/vacant land;
- control of advertisements/signs/hoardings/ 'clutter';
- enforcement of planning control including deterrence of unauthorised development.

Traffic Measures and Parking Provision

Apart from the major problems of regional inequalities, plus social and economic deprivation of many years' standing, probably the potentially most damaging factor in the modern-day environment is that of vehicular traffic. Clearly, this subject is so wide in itself that a selective approach is necessary here.

In the same way as the individual regards his or her personal house property almost as a sacrosanct object (with assumed inviolable rights both of protection from adjoining development activity, but at the same time invested in the owner's eyes with a prescriptive right of change and for expansion), so it is with the private motor vehicle and all the concomitant expectations of accessibility, road space and parking provision.

The whole question of uncontrolled urban motor traffic and car usage is heavily related to the fundamental matter of substainability (which we shall consider in Part III, in wider terms). An essential consideration in achieving urban quality of any given area must therefore be the question of degree of traffic accessibility. Additionally, there is the constant struggle between the voracious demands of the motor car and the less instantly appealing but much more economical provisions offered by public transport. The serious traffic congestion increasingly suffered by our town centres, main roads and residential areas both from moving or stationary traffic and from lack of adequate off-street public parking is a problem which at times threatens to engulf some areas indefinitely. One must add to this the damage to the environment which occurs from causes such as pollution, vibration, illicit parking, or the demands of new road construction in terms of the break-up of hitherto close-knit areas and the destruction of their character. This destruction may derive either from insensitive 'remedial' action or conversely from inaction (or

TOWN PLANNING FRAMEWORK

Figure 12 Traffic impact and relief: Epsom.
Urban experience is essentially for people. Wheeled vehicles in ancient Rome were soon forbidden to that city as it grew. Medieval times in Europe provided little surplus for the luxury of such useful encumbrance, and the velocity of personal mobility, through towns, only barely exceeded that of a man walking, when the horse-drawn carriage or cart choked narrow streets.

Today the urban populations of the world have demanded the convenience and the speed of the motor vehicle, in virtually every area of everyday life. The city has become the victim of the car, and its vital experience has been largely denied to people who are themselves victimised by pollution, and psychologically modified by the almost universal use of motor transport.

Epsom, Surrey: (left) High Street and entrance to (right) Ashley Shopping Centre.
Source: Michael Parfect

delayed action), where bypass roads could physically be provided, but where sufficient funds are not available in the face of too many other priorities. There is also the risk of alternative traffic reduction/calming measures having either a temporary or permanent effect of displacing people or traffic elsewhere, giving a false impression of overall neighbourhood benefit or improvement.

This is by no means an indictment of Highways Authorities; all too often they are required to deal with overall areas and problems that are disproportionately large, in terms of the human and financial resources available. Consequently, they are obliged to work to, first, a system of road-funding priorities, in progressive stages centring around a Transport Preparation Pool (TPP) for each present County area and, second, a system of central government grants where major roads are concerned. (The TPP 'prioritising' system, plus the Highways Authority's road improvement and construction programme, have been reflected in county structure plans and in district local plans, and are thus subject to public participation, comment and debate).

But the provision of new or improved circulation routes for traffic is not sufficient in itself to safeguard either the physical environment or, just as importantly, the convenience and safety of its users, whether vehicle-borne, cyclist or pedestrian. There is a fine balance to be struck between how far one discourages traffic from 'using' or flowing through a particular area, and how far one makes that area accessible to both vehicular movement and parking. Judicious introduction of 'resident parking schemes' to control unauthorised on-street parking by commuters, shoppers, etc. can bring relief to residential roads, by means of residents' parking permits, issued by the local councils (provided these do not militate against other valid parking needs). Then again one must consider to what extent one should instigate (and thus absolve) the effect of roads and parking provision by soft and hard landscaping, avoidance of unsightly 'left-over' sites and pieces of land, and the provision of general 'humanising' measures such as seats, urban sculpture, bollards, change of surface materials to demarcate different types of area, etc. In all of this, the needs of the pedestrian and the cyclist must be considered to be of equal importance to those of the vehicle. The many advantages of the cycling mode of personal transport, including benefits to the environment, are only now being properly recognised by government, and provided for by Highways Authorities.

Vehicular plans should be supplemented by well thought-out parking plans and traffic-calming measures (seen in an overall context) and augmented by cycle routes and pedestrian plans wherever appropriate or practicable. Frequently in the past consideration has been all about the provision of vehicular plans, as though movement of traffic (important as it is) were of greater importance than location, quantum and accessibility of parking provision in town centres, or could legitimately override the specific needs of pedestrian or cycle access and circulation. Nowadays, Highways Authority TPUs (Transportation Planning Units) are more conscious of, and adept in, relating traffic-calming, parking, cycling and pedestrian needs to the general planning of traffic movement. However, much is still left to under-funded local councils to try to provide in terms of these humanising environmental needs, including streetscape, planting and general public realm works.

Funding for such environmental works and for segregated pedestrian provision is not usually centrally provided or grant related, other than within 'accommodation' works which are immediately adjacent and related to new carriageway works (i.e. which essentially may be said to form part of the highway). The

Figure 13 Traffic-calming in Lewes.
Parking restrictions, speed zoning, and the physical reduction of excessive entry speeds to such areas, are becoming essential tools in lower-density residential parts of towns, subject to increasing use as short fast routes, for drivers finding other ways to work, leisure, and shopping. The road-narrowing 'chicane' favoured in France achieves similar effects over fast traffic, and the visual absorption of such devices in future planning is important, if road marking and over-signposting is to be avoided.
Source: Michael Parfect

rest normally has to come either from the local authority's coffers or sometimes is 'development-related', i.e. forming part of new development(s), 'planning gain' deals, and the like – a route which is not without its problems.

The same applies to public parking provision. The local authority, if it is lucky, may own some land in its town centre which can be used for car parking purposes, always assuming that these pieces of land are conveniently situated for people to gain access to the town centre facilities, and that the parking areas have fully practicable access from the primary traffic circulation. With the ever-increasing growth in traffic and parking demand, it will often be desirable or necessary for these ground-level parking areas (where of adequate extent and suitable location) to be developed into multi-storey car parks. However, the necessary funding for multi-storey parking structures (which are expensive items even without the requisite external 'finishing' treatment needed to make them acceptable elements of the urban scene) is very rarely within the financial resources of local authorities, who are often heavily restricted as to capital expenditure. Their provision once again has to be related to a development deal, frequently leading to yet more parking demand. (Nowadays, there are also issues of vandalism or crime affecting the question of provision of multi-storey car parks.)

At the bottom of this tangled and unsatisfactory situation lies a dilemma. Either traffic flees the town (it seems) or it swamps it – either one discourages to the maximum extent the vehicular use of central and other areas (including the thorny question of use of non-primary roads by essentially 'through' traffic), or one bends over backwards to accommodate all traffic potentially desirous of use, access or termination by spreading it out within general or particular areas. Quite often, when there is heavy growth of traffic both externally and internally, there are a limited number of options:

1. the construction of major new peripheral routes (only feasible where disruption of sensitive 'corridor' areas can be avoided, or where the implications of expenditure are not exorbitant);
2. the construction of inner relief roads which can also act as primary distributors for both local and through-traffic (where the latter is unavoidable); linked with environmental upgrading to areas thereby 'relieved' of traffic, e.g. by means of 'traffic-calming' measures;
3. the adaptation of existing routes, upgrading them for traffic, but avoiding downgrading them from an environmental viewpoint, by the use of traffic management techniques including forecasting of traffic flows via computer modelling of alternative assignment patterns. By this means, other routes such as shopping streets may be relieved of traffic pressures, where previously they had to take heavy flows (though the wisdom of this is not always apparent to local residents traditionally wedded to the idea of total freedom to drive their motor cars anywhere!);
4. the exclusion/banning of all motor traffic, other than emergency or service vehicles, for all or part of the day.

Both (1) and (2), however, carry penalties of environmental disturbance from the impact of new road construction. Unless particularly heavy flows of traffic (actual or forecast) have to be dealt with,[33] necessitating solutions such as (1) and (2) above, i.e. the construction of new urban bypasses or relief roads, the most pragmatic solution will usually be that of traffic management utilising existing main road networks (i.e. solution (3)). This may not only provide the possibility of deterring through-traffic from using minor roads as 'rat runs', but also enable the pedestrianisation of some areas, by the creation of cross-patterns of service roads and foot streets, as was done in various Continental towns such as Copenhagen and Munich as far back as the late 1950s/early 1960s, and which was followed up in Britain (albeit in a rather more limited way) notably in Norwich and Salisbury during the mid-1960s. But ultimately the accommodation of very large traffic volumes, which normally contain a great amount of through-traffic, had sometimes to be brought about by the provision of relief roads, in order to protect both central and residential areas, provided that the cost could be justified and the money found.

By 1996, however, the great 'car economy' of the Thatcher years was showing signs of expiry. It is now seen increasingly that by linking urban design where appropriate with traffic management, the need for new road construction can be minimised, existing primary road capacities maximised, and the impact of traffic and its demands upon local environments at the same time minimised, by a number of essentially commonsense measures, all involving restrictions and discipline as to vehicular use. It is axiomatic that the maximum degree of appropriate use should be made of existing primary routes, to avoid the unnecessary construction of new roads and the destruction or disruption of existing urban areas.

Particularly useful tools in this context are:-

1 discouragement or exclusion of non-local or through-traffic, including measures such as 'traffic-calming' on minor roads and weight restrictions for HGVs (heavy goods vehicles), though all or most of these need to be linked with the existence of some type of local peripheral or relief routes;

2 provision of adequate off-street parking, especially in Central Area peripheries or on approach routes, to prevent (or minimise) unauthorised on-street parking, though the need for on-street parking where feasible is equally valid, provided that this does not lead to environmental deterioration and conflict of interest locally;

3 refinement or revision of road junction design, again in order to maximise flow and use of road space (the capacity of any road network is only as great in practice as the capacity of its junctions);

4 lastly, and most importantly, the degree to which the use of public transport and cycling can be maximised, and the use of private cars minimised. The latter is highly significant for urban environmental quality. Despite the experienced benefits of American and also European public rapid transit systems in city centres,[34] sometimes linked with 'park and ride' provisions) it seems we are fatally and incurably wedded to the maximum use of our own personal private vehicular transport; with its presumption/expectation of the maximum degree of vehicular accessibility and accommodation. This is a trend that has become both insatiable and unreasoning in modern society, and unless adequately controlled, could have irreversibly damaging effects both upon the built environment and in terms of air pollution. All this is well enough known and yet still we continue on this destructive course. It would seem that only the most severe restrictions, linked with punitive measures as to the unbridled use of the private motor vehicle, will suffice if other measures for the control and management of traffic, and a sensible level of use and improvement of the road system, prove inadequate.

Development Phasing Control

Where a substantial scheme of development is to be permitted (e.g. an extensive shopping/commercial complex or a housing estate), it is normally in the interests of the surrounding inhabitants and the planning authority itself for the latter to reach agreement with the developer/applicant on some form of phasing of the building proposals (including the provision of roads and other utilities, or 'infrastructure'). This has not to date been encouraged by the DoE in terms of local plan policies, being seen as too restrictive. Yet it is normally sensible and advantageous to phase development in terms of effects upon local environment and amenities, as well as upon flow and deployment of resources. Agreements or arrangements of this nature should normally be the subject of prior discussion and negotiation between the developer and the local authority, in order to be reflected in the planning consent, though local authorities are debarred from writing land phasing policies into Local Plans for specific sites.

By this means, it may be possible to make a large or major development more palatable to both the electorate and the planning committee – provided, of course, that the principle of the use of the land or site for the type and content of the development in question is established and agreed clearly at 'officer level' at the outset.

Another aspect, however, which cannot normally be controlled by planning powers, is the routeing, frequency and timing of movements of construction vehicles relative to building operations on a given site. The noise and disruption to neighbouring property owners and residents can be considerable and can give rise to heated protest (often a sequel linked to an original unpopular grant of planning permission, whether expressly or on appeal). In this situation, the only practicable redress will normally be via the 'abatement of nuisance' provisions vested in the Environmental Health and Control Department of the local authority (provided that 'nuisance' can be proved, within the scope and meaning of the statutory provisions).

Control of Demolition

Especially damaging visually to the environment (and to some extent also in terms of effect on land values within the locality) are the incidences of demolition of property and/or of sites left derelict or vacant, for whatever reason. These fall under a number of headings, outlined below.

Building demolition and tree felling

The main areas of potential problem and objection in respect of the uncontrolled demolition of property, from the viewpoint of effect on urban quality and fabric, are:-

1 loss of buildings which are valuable in themselves, particularly from a historic or architectural aspect;
2 failure to provide for suitable replacement building(s), etc. on the site, with regard to the context of the surrounding area, and/or contribution made by the previous building to the urban scene;
3 creation of unsightly 'gap' sites, without any indication as to how the land will be used/developed in future, and how long this is likely to remain the case.

In all this, the emphasis must be on what measure of control exists under planning powers, to prevent the loss of the building, and/or to regulate the manner and form of its replacement, in the interest both of the site itself and the surrounding area. (Enforcement however can be difficult!)

As might be expected, the relevant controls are found to be strongest in respect of statutorily listed buildings. In addition, however, there are controls in respect of the demolition of unlisted buildings or structures within conservation areas. Finally, powers have now been introduced into the planning legislation to control the demolition of properties within residential areas.

There are also additional controls in respect of the felling, topping, lopping or wilful damage of trees in conservation areas, or which are protected by Tree Preservation Orders (TPOs). The local authority may well also require replanting in cases where tree felling is permitted (and where it has not). TPOs can be made either by the district or county council (usually the former, except where the county give the planning consent, or more than one district is involved), and can relate either to individual trees or groups of trees which are worthy of protection/retention. If a meritorious tree, which is not however covered by a TPO, is threatened, then a provisional Tree Preservation Order can be issued. This protects the tree from felling for a maximum of 6 months, pending the formal confirmation of the Order.

The basic provisions relating to control of demolition of listed buildings and for both listed and unlisted buildings in conservation areas have already been set out in the earlier section 'Conservation and Preservation' (p. 42). Now, even partial demolition of an unlisted building in a conservation area requires special 'Conservation Area consent', thus illustrating the degree to which the pendulum has now swung, in terms of protection of the environmental heritage via legislation and experience. Demolition of any building – or part of a building – in a conservation area in fact requires 'Conservation Area consent' unless already covered by another form of protection, i.e. as a statutory listed building or ancient monument. (There are certain minor exceptions to this very wide-ranging control of demolition, viz. buildings of less than 115 cubic metres in volume, or walls and fences, etc. less than 1 metre high alongside a road or footpath, or less than 2 metres high otherwise. Additionally, there are cases where buildings in conservation areas, or listed buildings in their own right, are required to be demolished by statutory order under other powers, e.g. in the event of dangerous dereliction or partial collapse of a building (under the Public Health Acts).

Derelict sites and buildings/vacant land

Two of the most common 'eyesore' situations, particularly in central urban areas, are either where sites contain unsightly buildings that are vacant and also fully or partly derelict, or where the sites are themselves left vacant and exposed, and become a depository for litter of various kinds and quantities. This is usually where the 'eyesore site' has come about through failure to provide for suitable replacement buildings on the site (in the case of both types of circumstance, indicated above), or where there is uncertainty or stalemate as to how the land should be used and/or developed in the future. The root causes for the site being vacant or neglected may differ, but are most often some form of either market forces or 'planning blight', or a combination of the two. Normally vacant or derelict land in town centres or residential areas (and often other land as well) is too valuable to be left lying around undeveloped for long – but even if the land is left in this state for a relatively short time, it soon becomes disruptive visually, or worse. A site left vacant or derelict in suburban Surrey, for example, will most likely be in that state owing to problems of 'development hope value' not being realised (i.e. where very restrictive planning policies are in force). In contrast, the vacant or derelict site in a depressed area of the North of England, Scotland or Wales will be simply one of many where economic life is either on the wane or has departed altogether, and where development finance cannot or will not be injected despite availability of planning consent.

Local authorities, as ever, find themselves largely confined to operating a range of negative powers, which may be either preventive or regulatory, mitigating or interventionist, but only occasionally punitive in nature (though the scope for the latter two reduces steadily in direct proportion to the encouragement of the 'market economy'). Councils generally are reluctant – or unable – to act where the situation demands compulsory purchase, or other forms of committal of scarce funding, but where at the same time they are uncertain as to reimbursement or adequate return financially for their intervention. This will apply equally to the purchase of, say, a derelict or neglected listed building, or of land which may in fact hold the key to a beneficial form of otherwise private development. It will usually be left to market forces, once again, to provide the necessary impetus or 'intervention' – but it will not happen if it is not expedient.

Nevertheless, left-over pieces of land (normally those that are redundant from the development aspect, such as odd corners and triangles of land remaining without vehicular access or development potential after an urban relief road has been driven through) can be upgraded in some cases by imaginative 'landscaping' with paving, seats, shrubs, climbing plants, etc. Action of this nature can ensure, for very little cost to the council, removal of existing problems of waste dumping and/or litter, and screening or embellishment of unsightly exposed end walls where properties have had to be removed, e.g. for road-widening purposes.

Alternatively, vacant sites awaiting development (i.e. as soon as planning consent is issued or 'market conditions' improve) can be effectively screened by tall hoardings, which can be attractively decorated with suitable paintwork areas and selectively designated areas for display of advertisements. These are temporary and piecemeal measures in essence, but are not an excuse for public or private sector inaction. If, however, an area is particularly run down, especially in inner-city locations, the local authority may be able to designate it as a priority for visual or other treatment – funds may be available for this type of situation from special government-backed programmes, where the case is sufficiently compelling in itself or the site particularly sensitive.

Cases in residential areas, where vacant land (awaiting development) becomes untidy, or where front or rear garden areas are misused by the occupants/owners of the houses in question for unauthorised storage of materials, or for repairs on motor vehicles (sometimes involving the additional presence of derelict vehicles for 'cannibalisation' purposes) can be relatively simply dealt with by the local council, under planning powers, though even this can be time-consuming for hard-pressed staff of planning departments, since evidence has to be accumulated over a meaningful period of time.

More intractable, are cases where listed buildings are deliberately neglected (or worse) in the hope that they will become so run-down or derelict that the owner (or prospective owner), often a developer, will secure a planning permission for a new use (or new building) which would not normally have been forthcoming otherwise. In this circumstance, the local planning authority can itself take steps to repair the building, and recoup the cost of doing so from the owner, or it can even compulsorily purchase the property, repair or reinstate it, and sell it on, via the open market. However, many local authorities (particularly small authorities, with very limited financial resources) are notoriously reluctant to 'stick their necks out' financially in this way, in case they are unable to recoup the cost (owing to legal or other difficulties). They are thus more inclined to capitulate to the developer by allowing an office or other commercial use to ensure the future of the building (and get themselves off the financial hook at the same time). Much will depend, of course, on the nature and size of the local authority in question – the City of Chester, for example, has a very successful record of acquiring/repairing/renovating many historic buildings in its area, and of going on to do yet more. The 'rolling fund' system of purchase, reinstatement, and sale of old buildings worth preserving has frequently been proved to be highly rewarding and successful, especially when assisted by matching funds or subsidy by other bodies, including the DoE/English Heritage, county councils where applicable, and of course commercial bodies both with beneficent instincts and an eye to investment via 'public good works' linked to economic return. *Again, however, public finance is strictly limited in this respect, and thus more onus must fall on the private sector.*

Control of Advertisements and Signs

This aspect of urban quality merits attention, from the viewpoint of its visual effect upon the street scene generally. The Advertisement Regulations (which from time to time are revised), govern what sizes and types of advertisements can be erected adjoining the highway, or on fronts of buildings, with or without express consent under the Regulations). Some advertisements were by 1990 exempt from consent, or had 'deemed consent', e.g.

- professional nameplates, warning notices, direction signs or house name signs (up to 0.3 square metres in area);

- notices on churches, hotels, pubs, and certain other public buildings (up to 1.2 square metres in area);
- notices of meetings, etc. for charitable purposes (0.6 square metres maximum);
- election posters (must be removed 14 days after polling);
- most neighbourhood watch scheme signs; and
- temporary advertisements for sale/lease of residential property (up to 0.5 square metres in area) – though most local authorities will discourage proliferation of these, and require multiple signs to be combined into one board (under the relevant legislation).

The great majority of advertisements and signs in public areas are thankfully controllable, and many councils have local policies specifically to combat the adverse effects of sporadic signing or other forms of advertisement 'clutter', which can be of a disruptive intensity. There is no doubt about how severe and damaging that disruption would potentially be, were it not for local controls. Advertisements can, if well designed and regulated, contribute acceptably to urban quality, in a way that is sympathetic (above all) to the particular urban context.

Nuisances

Sundry matters that affect urban quality in different ways also need to be examined and noted. They can be taken in two broad groups, namely (a) Flyposting and litter, and (b) Noise, smells, and nuisance/noxious activities.

Flyposting and litter

Flyposting might be said to be a cross between unauthorised advertising and a type of litter deposit. However one regards it, the effects of flyposting are usually detrimental to any urban area, especially where other forms of uncontrolled advertising have taken place. This is yet another thorn in the flesh of local authorities already fully occupied with trying to exert control over the environment. Whilst matters such as substandard or unauthorised shopfront advertisements or signs can normally be dealt with fairly effectively (given adequate planning control staff), since the responsible owner of the land or property can usually be traced and formally approached to rectify the matter, flyposting is a rather different problem by its very nature. Much unauthorised affixing of temporary notices advertising forthcoming events (be they car boot sales, heavy metal/rock sessions, stock car racing, or even jumble sales) are organised and also advertised by non-local commercial and other bodies, over whom the local council has no immediate jurisdiction. The most that the council can do in this situation is first to try to contact the relevant external body that has created the flyposting, and request them to remove it within a certain period, or failing such action, it can take its own steps to remove the offending signs. *However, very few local authorities nowadays can afford the staff time (or indeed possess the requisite staffing resources) regularly to patrol their area, identifying and removing illicit signs from lampstandards, hoardings, or buildings – it is now simply uneconomic to do so, whatever objections it may raise.* Scarce planning and enforcement staff need to be deployed on more weighty and pressing matters, and unfortunately this fact has to be recognised.

Litter is an even more serious problem in many an urban area, though it does not fall within the remit of planning departments, being traditionally the problem of the City or Borough Engineer's staff. Many district councils in the late 1980s/early 1990s combined their planning, building control and engineering functions into a Technical Services Department, or the like. This makes sense in terms of grouping together various functions for the physical control and improvement of the environment.

Like any other function for the improvement or maintenance of urban quality though, adequate funding must be provided to ensure an adequate standard of service. Compulsory competitive tendering (CCT) has meant that many services previously provided by local authorities are now carried out by private firms of contractors – though whether for better or otherwise is very much a matter of individual circumstance or judgement; experience has shown that widespread privatisation is not a universal recipe for success, especially where 'value for money' is concerned.

There is no doubt that, during the 1980s and early 1990s, the problem of litter in many of Britain's urban and city areas became noticeably worse. Comparisons with other West European countries are not encouraging in this respect. The City of Paris, for instance, is reputed to spend around 10 times as much in litter control as does London, and has invested more extensively in modern equipment and technology for dealing with litter and dog faeces than has the British capital. A visual inspection of comparative public areas, in both cases, may well be enough to bear out the effects of the two different levels of investment, in this 'tale of two cities'.

Noise nuisance/traffic pollution

We have examined various forms of visual nuisance, such as litter and derelict land and buildings. *Bad as these often are, other forms of urban intrusion, specifically noise, fumes and smells can sometimes be worse.* Traffic is normally the major source of the latter types of nuisance; but other 20th-century activities such as car-breaking, 'non-conforming' industries, fish-and-chip shops and other food establishments can equally well pose problems, especially when insensitively sited in the urban context, or where inadequate controls on nuisance exist. Nowadays, Environmental Health Regulations are so demanding, and control is so tight, that local authorities are able to keep many existing or incipient nuisances in check fairly effectively. The main exceptions to this are: bad planning (or engineering) or generally lack of forethought in permitting an industrial type of activity; building a new road without sufficient care for the locality; or where the local authority has to deal with unsocial activities enjoying existing use rights, albeit often in situations that do not conform with the town's development plan (i.e. 'non-conforming uses', though not all of these are necessarily so detrimental as to require removal or relocation). This latter category had, in fact, prior to the mid-1980s, been the subject of a general 'winkling out' process by local councils, coupled where feasible with relocation of the non-conforming industries in question out of assorted areas and into new purpose-built industrial areas. The underlying difficulty was most often one of economics – rentals and land values for backyard or backstreet sites are usually way below those for new, fully serviced, segregated industrial estates or areas. This difficulty was then exacerbated by a central government directive (inspired by Thatcherite 'economic determination'), which decreed that both existing and prospective 'non-conforming' uses should be allowed (and even encouraged) to remain or set up in residential areas, in places 'where no material harm would result', or where 'nuisance and noxious activities could nevertheless be controlled'.

But by far the greatest potential source of nuisance is, as already stated, that arising from traffic both in its existing levels and in its exponential levels of growth. This is manifested chiefly in terms of noise and fumes, but the additional elements of inconvenience, physical damage and erosion, and in many cases, of personal danger, are not inconsiderable.

The growth of road traffic in Britain is of course quite external to local authority control, being influenced chiefly by economic factors such as population level, wealth distribution, travel times and personal predilections. Central government does, however, play some role, in that it can influence mode of travel by its attitude to public transport (particularly railways), and more importantly by its policy towards subsidising or encouraging financial investment in the latter. The record has shown during the 1990s and earlier a lack of support for railway subsidisation commensurate with the government plans for rail privatisation.

In this situation, the problem of reducing noise, fumes and other urban pollution aspects of increased road usage depends for its mitigation upon the skills and resources that hard-pressed Highways Authorities can bring to bear. There is clearly a limit on how far the problem can be mitigated (as opposed to being solved), but undoubtedly this specific issue has as many implications for urban quality in its widest sense as do land values and ownership patterns for the planning system generally, as mentioned at the outset of this study.

Enforcement of Planning Control, Including Deterrence of Unauthorised Development

An increasingly frequent problem in the late 20th-century urban scene is that of 'unauthorised development' of land or buildings, or of other breaches of planning control, and how the local authority is to deal with it. As has been pointed out earlier, not all development requires planning consent (i.e. those works that come within the statutory definition of 'permitted development' are exempt from the need for express approval), but the great majority of constructional works, changes of use and appearance, or proposed use of undeveloped land do require consent. (Most structural works also need building regulation approval, which is an entirely separate system with its own particular implications for urban quality).[35]

Except in times of recession and of consequent slump in constructional activities generally, when the numbers of planning and building control applications fall away to some extent, the priority of dealing with the constant influx of new applications means that there is very little spare capacity in terms of development control staff time to go around the whole of their urban area picking up cases of unauthorised development. If the local planning authority does happen to spot such cases itself, it is most often via the circulation of their Building Control Officers (who need to spend less time checking plans in-house than inspecting various works on-site).

In many instances, however, it will be 'neighbour notification', in different forms, which draws the local authority's attention to the relevant breach of planning control. Local resident vigilance is rarely found to be lacking in this respect! Not all alleged cases of breach of control are capable of confirmation in planning control terms, in the event — but usually there is no smoke without fire. In terms of urban quality, a single domestic extension, shopfront and/or advertisement sign may have an unfortunate effect, when both lacking the necessary consent and design input. A multiplicity of cases of 'unauthorised development' is almost certain to do so. A single case, however, of unauthorised development or alteration in a conservation area, or on a listed building, can have a highly undesirable, even disastrous effect. The celebrated case of the non-conforming window-blind in the Royal Crescent at Bath — a Grade I group of buildings of international importance — brought 'architectural enforcement' right into the public eye, and was one instance where the DoE upheld the local authority's ruling that the offending element should be removed. (Some might say, however, that the closing down of a car-breaking activity in their neighbour's rear garden was just as desirable a course of action!)

Before going further into what types of powers local planning authorities have in terms of dealing with breaches of control, mention should be made of two particular aspects directly affecting enforcement action generally, as follows: -

- Appeal against enforcement;
- 'criminalisation' of planning offences: the Carnwath Report.

Appeal against enforcement

It should not be overlooked, when considering (or faced with) enforcement action authorised by the council, that the private individual or body being proceeded against has a right of appeal. The exercise of this right has the effect of halting any physical process of rectifying the works, or returning to the 'status quo' for several months usually, whilst the propriety or validity of the proposed enforcement action is considered by higher authority (normally the DoE). Frustrating as this exercise of private rights may be to an aggrieved third party or a local council intent upon rectifying a perceived wrong, it is all part of our democratic system of legal checks and balances.

'Criminalisation' of planning offences: the Carnwath Report

The increasing frequency of 'unauthorised development' in the 1980s and 1990s, as experienced by the majority of planning authorities, led the District Planning Officers' Society to press the government for the 'criminalisation' of (deliberate) planning contravention. The basis for this line of argument was that, not only were the effects on the environment harmful and often serious, but also that to fail to deal severely with such misdemeanours (or to turn a blind eye) was inequitable to the majority of law-abiding citizens, and in general was a failure to ensure that 'the punishment fits the crime' (as W.S. Gilbert would have said). There was no doubt in the Chief Planning Officers' minds that the overt or covert flouting of planning control, deliberate or otherwise, *was* a crime, to be dealt with accordingly. The argument gained ground and an in-depth report was made to the Secretary of State for the Environment by Robert Carnwath QC (hereinafter referred to as 'the Carnwath Report'), and was published in February 1989. In general, the Carnwath Report supported the case for 'criminalisation' of planning contraventions, and made various recommendations, a number of which were adopted by the Secretary of State and found their way into new planning legislation, in particular the Town and Country Planning Act of 1990 and the Planning and Compensation Act of 1991 (the latter substantially strengthening and improving the relevant provisions of Part VII of the 1990 Act). The main features of the latter legislation as at 25 July 1991 (the date of the Royal Assent to the Planning and Compensation Act) were as follows.

1 The new enforcement regime – the new improved enforcement powers of the 1991 Act related to:

- *the power to serve a 'Planning Contravention Notice'* where it appears that there may have been a breach of planning control and the Planning Authority require information about activities on the land, or the nature of the recipients' interest in the land (new Section 171C of the 1990 Act);
- *the power to serve a 'Breach of Condition Notice'* where there is failure to comply with any condition or limitation imposed on a grant of planning permission (new Section 187A of 1990 Act);
- *the ability to seek an injunction, in the High Court or County Court,* to restrain any actual or apprehended breach of planning control (new Section 187B of 1990 Act);
- *power to serve a 'Stop Notice' to prohibit the use of land as the site for a caravan* occupied as a person's only or main residence, and to make a stop notice immediately effective where the authority consider it justified

(amended Sections 183 and 184 of the 1990 Act); and

- *improved powers of entry on to land* for the planning authority's authorised officer to obtain information required for enforcement purposes.

2 Enhanced penalties for certain enforcement offences
New penalty provisions were introduced, so that in future a maximum fine of £20,000 would apply for each of the following offences:

- failure to comply with the requirements of a Planning Enforcement Notice, as per Section 179(8) of the amended 1990 Act;
- contravention of the prohibition in a stop notice, in accordance with Section 187(2) of the 1990 Act as amended;
- contravention of the provisions of the control of listed buildings, in accordance with Section 9(4) of the Planning (Listed Buildings and Conservation Areas) Act 1990;
- failure to comply with the requirements of an enforcement notice relating to a listed building, or conservation area consent, as per Section 43(5) of the Listed Buildings Act 1990; and
- failure to comply with the requirements of a Tree Preservation Order, as per Section 210(2) of the 1990 Act.

The increased penalties were presented as consistent with government policy stated in the White Paper entitled 'Crime, Justice and Protecting the Public', of February 1990. The point was also made that, in punishing offences entailing a risk to public health and safety, creating a public nuisance, or harming the environment, the penalties should take account of the resulting profits or savings accruing to the offending person or body.

Coming in a period of government when market forces ruled the day in almost everything, and consequently when private individuals and developers alike were encouraged to exercise the greatest amount of 'enterprise' in every field (with the local authority cast simply in the role of 'enablers'), it is perhaps surprising and certainly refreshing to find that 'planning offenders' should suddenly themselves be subject to such penalties and measures for restoring control. Planning should not of course be merely restrictive or punitive, but should also be flexible and creative. Nevertheless, it should at all times have teeth to deal with those who would flout it.

Postscript: Death of a Planning Officer On 20 June 1991, Harry Collinson, a principal Planning Officer of Derwentside District Council in County Durham, was shot and killed by the owner of a bungalow who had built it without planning permission in open countryside, close to the A68 road to Scotland. Mr Collinson was supervising its demolition, ordered by the district council in culmination of an enforcement action it had been obliged to pursue, when he was killed. The owner was subsequently jailed for life. The case, extreme as it is, illustrates 'one of the most contentious issues in modern-day planning' (as a Sunday newspaper later put it): the conflict between 'the determination of [some] individuals to live according to their own philosophy' and a situation where the environment is protected by law against illicit development. It highlights the antipathy which can be felt by some people for (the officers of) local planning authorities who conscientiously enforce the law in respect of the environment. But when no alternatives or 'room for manoeuvre' exist, as in the above case, no defence exists for those who act in defiance of the law, to whatever degree.

URBAN ENHANCEMENT MEASURES AND PROVISIONS

Galbraith: 'Private Affluence and Public Squalor' Dictum

It was as long ago as 1958 that the eminent economist and US presidential adviser, Professor John Kenneth Galbraith, first spoke of 'private affluence and public squalor'.

The 1930s New Deal in the US of President Roosevelt, and the 1940s Welfare State in Britain, introduced by the first post-war Labour government (on the back of the New Deal) both 'softened the inhumanities of market economics and helped create an electoral majority of the broadly contented and self-satisfied'. This majority, plus the politicians who later took up and furthered its mood, came to define and control modern democratic politics through the 1980s and beyond. These right-wing politicians (for such they were) 'had no respect or understanding – rather enormous contempt and hostility – for the public policies and spending that allowed them to become contented and self-satisfied in the first place' (and no compassion for those left outside their contentment). In his 1992 book *The Culture of Contentment*, Professor Galbraith predicts that 'this myopic and self-destructive phase is not likely to end soon', and indicates no willingness (either in the US or Britain) to return to the policies of 'renewed compassion, and public investment paid for by taxes'.

Galbraith's own political convictions appear in this book as clearly defined as ever, i.e. that 'unlimited, unguided private activity is as destructive as unlimited socialist public control and interference'. 'The miracle of the modern mixed economy (capitalism with substantial public investment and guidance) between the Thirties and the Seventies, saved the Western world from Marxism', says Galbraith – 'But it unexpectedly also contained the seeds of its own gradual shabbification', whereby the role, power (particularly fiscal strength), and capability for intervention of the public sector since the 1970s has been progressively and intentionally eroded. Those more comfortably placed socio-economic groups who (through the appropriate political machinery) have encouraged or connived at the severe restraints upon the public sector, are increasingly beginning to regret the consequent deterioration of the public environment.

Private and Public Sector Joint Action Potential

The more thoughtful amongst them will also have been reflecting on the legacy of shabbiness which the manifest neglect of the public realm will be leaving for future generations. And this is not all. The economic recession which hit Britain, like other countries, in the early 1990s has in turn begun to cause an inversion effect, whereby the previously highly affluent private sector can no longer afford to carry out or subsidise works of environmental improvement, and is looking to the public authorities to play a greater part in making those provisions for society. We note this especially.

Clearly, the answers for the future will lie increasingly in the potential for joint action (and funding) between the private and public sectors, and in particular the parallel need for involvement of the local business communities as well as the general public in measures for urban/environmental enhancement. (It may be pointed out here that local authorities in fact have been dependent on the 'package deals' of private developers since the late 1960s, when public sector capital expenditure had to reduce.)

To enable this to take place successfully, a whole range of antagonisms, prejudices, misinformations and misunderstandings (some of them deep-rooted) concerning the environment, civic design, architecture and the role of planning in the public sector needs to be overturned. Many of these prejudices and/or antagonisms, in the minds of both the private business sector and the public generally, stem from the heady days of the 1960s. In those days, local authority planning and involvement in development on a wide scale tended to reign supreme, with the private sector occupying second place (and often made to feel it). In keeping with the 1964–70 period of Labour Party government, local authorities also were to acquire a 'socialistic' character, which manifested itself both politically and in terms of environmental design and planning control. There was a combined image of Le Corbusier and Gaitskellism in the Britain of the 1960s, and the flavour of that image has dogged local authorities ever since in the eyes of the public, to a damaging extent. 'For Planning, read Socialism' was the impression of many, especially in the southern half of the country, now however almost entirely coloured gold or red on the local authority political map of Britain. (There were in fact few Labour–Conservative differences during this era regarding mass housing development.)

The 1960s, originally so optimistically shining with bright ideas about new community living, with tower and slab blocks to house the expanding residential and office populations, upper-level walkways to segregate pedestrians from traffic, pedestrian precincts and neighbourhood shopping parades wherever possible, multi-storey car parks likewise, and communal green areas between the new vertical blocks, gradually came to acquire a tarnished image, as indeed its buildings did, as they were neglected or became outdated, and are easily identifiable as such today.

The notion of Comprehensive Development Areas, also stemming from the 1960s, in due course became discredited, as they were too closely identified with 'Modernistic' and 'monolithic' development – hardly ever a concept to sit happily in the image of traditionally picturesque and informally planned Britain (leaving aside, of course, the earlier more elegant and successful examples in that field, such as Bath, Brighton, Cheltenham and Edinburgh New Town of 200 years ago, the difference being that historic architecture always scores more points than its more modern counterpart). Prejudice always dies hard, and it is still around, where planning, architecture and the role of the public sector are concerned. As ever, though, the record is mixed – there is a mixture of unfounded antagonism, but also of legitimate criticism in certain areas such as tower block living, where the Ronan Point collapse, crime and juvenile delinquency have sounded the death-knell to what was basically a political mistake of applying too widely a formula for instant large-scale housing involving the minimum of land. It was dressed up at the time by the politicians of all parties as a brave new step forward in social housing at high densities, and the buck deftly passed subsequently to the architects in a charge which still unfairly sticks.

Equally, if there is a jaundiced view of the legacy of 1960s planning and architecture for its reflection of now discredited 'international' values and notions, there is probably also a growing disaffection with private development sector preoccupation with capital-intensive 'prestige' projects. These often would appear to be directed mainly to maximum economic return rather than to meeting prevalent social and environmental needs, and have given rise to an elitist 'inner circle' of big-city architects interested only in such major projects. It is noteworthy that the more advanced (and to some extent extreme) forms of new technology in building are most often directly related to those prestigious 'one-off' schemes, for which the maximum level of capital for investment in building is available – as against the more 'everyday' types of situation where building and design skills are needed to improve, enhance or in other ways contribute integrally to the environment. It is a matter for public speculation as to whether the relevant professional interest lies mainly in the specialist schemes themselves, the application of new technology, the financial resources and rewards, or in all of these. (Housebuilding in Britain, however, presents a curious anomaly in respect of the private sector's continuing fixation with pastiche historical styles still deemed necessary to satisfy the market.)

The arguments and prejudices surrounding the opposing public and private sector roles, their merits and demerits, benefits and disbenefits, and so on are endless and eventually lead nowhere. The important and urgent need is to resolve, or at least put aside, as many of these as possible, in the search for common ground in the really essential business of improving our existing urban areas. Setting new, soundly-based and generally applicable standards of planning framework and building design to cater for the various ills and deficiencies of our environment, whilst preserving our past heritage as best we can, is what is needed. At the same time we have to travel sensibly, rather than rashly, into the future.

In the realism of today's situation, where land and property ownerships have become increasingly fragmented, where neither the private development sector nor the public authority realm has overriding power to shape environments on a wide scale; but where the power of land values and available monetary investment will ultimately be the most persuasive factor, the common ground for urban quality improvement on anything but a purely individual basis must be carefully examined, for what potential still exists.

The most fruitful ground for progress towards a revival of civic pride would be to engender a consensus view of the wisdom of private property/land value enhancement, linked with public sector and business community joint action on the need for environmental improvement generally. The latter would include such aspects as street 'facelift' schemes; street furniture renewal/provision; 'percent for art' benefits; tree planting where appropriate; and, most importantly, traffic restraint/exclusion wherever possible. Preservation/conservation will also have a significant role to play. The considerable potential benefits of tourism (properly managed and provided for) should also be taken into account. Not all the necessary capital resources should come from new development, via 'planning gain' or 'percent for art', valuable as both these mechanisms may be: money for some essential needs of street or building improvement should also be made available from local or central government resources, though of course in a period of continuing recession this will be no easy matter. In some special circumstances, there may be a case for European Union money also to be made available; but here the government's record until recently of automatically reducing the local authority's overall borrowing limit by the amount of the EU grant represents a shabby, underhand practice, which compares poorly with fiscal practice in other West European countries. (It also does little to commend us to Brussels, though the situation is now at least partially resolved.)

Property/Land Value Enhancement

In any consideration of urban enhancement, the question of value enhancement of individual sites or buildings will be an underlying factor, and will probably be the first consideration of most individual owners. There are basically three ways in which property values may be thus enhanced:

1 increased value of a site or building reflected as part of a local environmental improvement, e.g. a street scheme or a pedestrianised precinct;
2 increased value of a site by virtue of planning consent for a building development upon it; or for a floorspace increase, or of change to a

more lucrative form of use, via the requisite planning consent; and/or
3 enhanced value of a development site by virtue of a more attractive/intelligent/skilful and commodious form of design of the building(s) for that site.

Of these three factors, the owner or developer is likely to be most interested in (2). It will be a matter of either good fortune or the farsightedness of himself, his architect (if he has one) or the planning authority if he is interested also in (1) and (3). It will, however, be factors (1) or (3) above which will hold out the greatest chance of an improvement in urban quality.

It is (or should be) axiomatic that 'good design equals good investment' in whatever context, quantity or scale this principle is applied. It is so obvious that it barely needs discussion – yet there are still some would-be developers who fail to recognise the wisdom of employing a skilled designer to put forward and subsequently carry out their schemes. Cities are positively littered with examples of office or other commercial buildings to which the minimum amount of design thought has been given, which employ the most utilitarian materials and detailing, and which thus present the most dreary and bland of images.

Fragmented land ownership patterns, as is often the case in Britain, usually have some bearing on incoherent forms of development. This pattern will persist particularly in a recession when developers lack the money to buy up complete packages of land, and when local authorities are left impotent to assist them to any marked extent via local intervention. Even when the developer may own all the land he requires to carry out a scheme for a new housing estate (irrespective of size), he may nevertheless run into problems with the local authority in terms of getting planning consent, either in outline or detailed form. This is more likely to happen in cases where the builder-developer has decided not to employ the design and planning skills of a competent firm of architects and planners, and in this event he will have nobody to blame but himself when the requisite consent is not granted. Alternatively, where the local authority does not discriminate between good, bad or indifferent design (happily there is a decreasing number of these), the results in terms of buildings on the ground may be disappointing.

With these particular problems in mind, Essex County Planning Department urban design section, under Melvin Dunbar, produced during the 1970s its own *Design Guide for Residential Areas*, in which various forms of housing layout and design in tune with the Essex building vernacular style were set out in an attractive and well-considered way, with the objective of assisting local developers to produce satisfactory schemes for planning consent and implementation. The design guide was however intended neither as a comprehensive document, precluding flexibility or inventiveness, nor as a 'bible' for use all over the country, from Penzance to Peterborough (by both local authorities and developers alike) as some sort of instant safe formula for universal application. Nor was it intended to stifle or supplant the architect's role in designing and carrying out housing developments. Nevertheless, virtually all the above pitfalls have been duly fallen into (with architects being the most vociferous complainants of erosion of their rights). Yet the Essex Design Guide, for all that, marked a significant milestone in urban design in this country, and still stands the test of time when properly used. It was followed in the early 1980s by the Surrey Design Guide *Layout in Residential Areas*. This married in more specific detail the potentially conflicting needs of highway geometry and parking requirements with the considerations of amenity in housing layout, in a very successful and flexible way, owing largely to the fact that the Guide was produced by a joint working party composed both of Surrey District Planning Officer representatives and County

Planning and County Highway Department officers. (A further relevant aspect is the need nowadays to plan the layout of housing estates in particular for maximum protection against crime, such as burglary, assault and vandalism. Many local police forces will happily provide advice in this connection to planning authorities, both in the form of explanatory talks and via consultation on applications.)

Urban quality more often than not relates to variety and mixture of uses, in terms of creating a friendly environment. The impression should be avoided that urban quality is mainly to be obtained by means of monolithic land-use planning, i.e. via inflexible land-use zoning or major single ownerships. Too often in modern times such an approach has resulted in monotonous or soulless tracts of commercial or housing development, especially where combined with unimaginative or unsympathetic architecture. Both the public and private sectors have been guilty of this approach, i.e. where local authorities have put too much faith in a 'clinical' or simplistic approach to land-use planning (tidy minds at work), and where the funding bodies behind developers have resisted the notion of mixed-use development as being more costly or difficult to finance than single-use schemes. The deadening results of this philosophy are still to be seen in our cities and towns, dating from the immediate post-war period of the 1950s and 1960s onward. Patterns of land ownership are highly relevant in this. Single ownerships can, however, facilitate the realisation of development schemes, whilst not precluding mixed-use proposals. Single-use development of a manageable and 'friendly' nature, such as in shopping centre schemes, can also produce urban quality if the content is not overwhelming or 'synthetic'.

Multiple-ownership patterns, whilst normally offering more chance of a vibrant and interesting built environment (if related to a coherent planning concept rather than merely left to the free play of market forces alone) can otherwise give greater rise to the messy, uneconomic and generally unsatisfactory process of 'development by appeal', which has been the cause of so many fragmented environments, lacking in any cohesive urban quality.

Preservation Pays

In any review of urban enhancement measures and of urban quality, the question of building preservation demands almost equal attention to that of land supply and utilisation. Our architectural heritage should be seen as belonging to the same category of 'dwindling commodity' as that of building land. There has been a growing realisation in recent times that the loss of historic buildings, either through neglect, decay, damage, official consent for demolition, or vandalism (including unauthorised demolition), so prevalent in the 1960s, cannot continue. The individual and collective effects upon the national architectural heritage of a failure to stem the tide are only too evident and painful. *Yet a steady depletion of the stock of historic buildings continues.*

This highlights and underlines the case for reuse of old buildings however and whenever this can be achieved. It is a subject deserving of more attention than it in fact receives, though an excellent lead in this direction was set in the late 1970s/early 1980s by a joint working party of the Historic Buildings Council for England and the British Tourist Authority which materialised in the BTA's publication *Britain's Historic Buildings: A Policy for their Future Use.* Many of its findings and recommendations remain valid today. There are, for instance, many urban areas where historic buildings are prone to deliberate arson attack because they actually have negative value to their owners.[36]

There are in principle various benefits to be gained from a policy of preservation and reuse of old buildings, including:-

1. a benefit to the environment in general, in terms of preserving our architectural heritage on its own merits, 'for the beauty, the character and the history it unquestionably represents and embodies' (as the BTA report put it);
2. an economic encouragement and enhanced financial return, via the retention and reuse of old buildings for commercial, residential and, in particular, tourism purposes (where both the direct and indirect use benefits in monetary terms are normally very considerable);
3. material savings in terms of prolonged life of existing building fabric, and the consequent reduced demands upon new or replacement materials; and the corresponding energy savings associated with avoidance of new construction works;
4. the 'boosting' effect of building conservation upon whole areas or neighbourhoods, on an accumulative basis, compared to the replacement of numbers of old properties by large new commercial blocks (which as often as not remain empty for long periods of time, sometimes ironically needing conversion back to housing in order to give them fresh economic life);
5. in the case of commercial development, congenial and prestige accommodation can alternatively be provided, often in central locations, e.g. for office uses, by the individual (or collective) reuse of older properties;
6. in the case of housing development, extra amounts of residential floorspace can be provided, often of an attractive character, by the conversion or subdivision of old buildings;
7. in the case of tourism needs and responses, a wide potential exists for the provision of attractive and interesting historic buildings or areas to see, stay in, eat or drink in, or shop in; and finally
8. reuse of certain old buildings for housing the homeless or socially/economically deprived would certainly help to meet an acute need, provided adequate safeguards are built in for the physical upkeep of the fabric and maintenance of its character.

In all the above instances, there will be an accompanying requirement not only for satisfactory maintenance and inherently good constructional and safety levels, but also for adequate access, parking, layout and planning standards generally. *At the same time, however, local authorities should avoid being too stringent or demanding in terms of applying modern 'health and safety' requirements (normally intended for new buildings) to existing older buildings.* The inflexible application of modern standards can mitigate against or stifle well-conceived conservation schemes. Either waivers of regulations or alternative structural options may well be necessary, especially in the case of Environmental Health and Building Control demands. Probably the most cogent observation to make is that, wherever possible and appropriate, old buildings should continue to be used for their original purpose, or for a use closely approximating to the original. There will, of course, be numerous exceptions to this, but the keynote to successful conservation will be sensitivity to both the context and the design of each historic building in question, leading to an informed flexibility of use where necessary. Financial resources and their availability will naturally continue to underpin any historic building preservation; *but the 'make or break factor' will still be the individual and/or political belief in the proven merits of conservation.*

Environmental Improvement via Planning Gain; 'Percent for Art'; Street Schemes; Grant Aiding; and Pedestrianisation Schemes

Enlarging more widely the above general considerations relating to individual site or building enhancement, whether in terms of values, design or preservation, there are various specific aspects which may also be applicable, depending on cases or circumstances. These are set out in the table below:

Type of scheme	Basis of benefit[37]
Planning gain	Individual schemes (development)
'Percent for art'	Individual schemes (buildings)
Street ('facelift') schemes	Individual schemes (buildings and street)
Grant aiding	Scope varies widely (with type of grant)
Pedestrianisation area schemes	Street or area basis
Traffic-calming and relief	Street or area basis
Landscaping/planting	Site or area basis

One specific aspect not to be overlooked in any consideration of urban quality is that of the adequacy or otherwise of services infrastructure, e.g. drainage and water services in particular, since many of these date from Victorian times and suffer from lack of maintenance/funding. What is below the surface is just as relevant as what is above ground!

Most important of all, however, is the need for the skills of town management to be applied to matters of urban quality concepts and provision, via structural and phased/costed programmes of action, avoiding the minimal 'patch and paste' reactive approach. Clearly, the local authority is the body that should be best placed to exercise the key control role of town management in respect of the provision of urban quality.

It is still open though to enterprising local authorities to take individual steps under any of the above headings (as set out in the table) which may appear relevant to its problems on the ground. Financial ways and means do exist to help in this process, and these are summarised below, relative to each type of urban improvement activity examined.

Planning gain[38]

It has to be emphasised that 'planning gain' must comply with the test of legality in terms of the environmental improvements sought being both directly related to and necessitated by the parent development in question. Examples of this might be landscaping improvements to the external setting around the site of the proposed new development, or improvements to the peripheral road system, at least as far as is essentially required to absorb the new development safely and commodiously in its urban context. *The amount of money involved will be for agreement between the local authority and the developer, but in no sense should the local authority be seen to be 'selling' planning permissions – an illegal practice of course. Legality must always remain the keynote in any 'planning gain'.*

'Percent for art'

A mechanism increasingly used in the US and much of mainland Europe, but relatively little known or applied in Britain, is the concept of 'percent for art'. Basically this consists of the developers of urban projects agreeing to donate a sum equalling 1% of the capital value of each project for the commissioning and provision of public works of urban art. (1.33% is levied by some local or city authorities, to cover administrative costs as well.) The works of art in question may take different forms, but paving and/or sculpture of various kinds is a fairly frequent option. External (or internal) decoration to buildings is another type of option, particularly relevant in modern times where so many new buildings or structures tend to lack anything much in the way of decorative architectural treatment.

As with 'planning gain', which is a related subject, there are questions of ensuring legality of 'percent for art' requirements, and of other aspects such as how a policy for 'percent for art' should be both worded and implemented, including budgetary matters of scope and methods of application, on and off site, etc. (The Arts Council of Great Britain produced a helpful publication: *Percent for Art – a Review* in the early 1990s which may be referred to for detailed information.)

'Percent for art' in a political context also provides local authorities with an extra opportunity to regain some of the influence they have lost (or have been deprived of) in the overt diminution of the public realm since the end of the 1970s – a concept which has now to be painfully rebuilt.

Street ('facelift') schemes

Moving from enhancements on the basis of benefits to individual buildings or sites (whether by good design of new development, preservation/renovation of existing buildings, or artistic embellishment of new buildings and occasionally their settings), we come to urban enhancement on a street or area basis. We first look at the street 'facelift' scheme, as it is popularly labelled.

One of the very first, and probably the best-known 'facelift' scheme in Britain was in Magdalen Street, Norwich, carried out in the early 1960s by the Civic Trust, under the co-ordination of the architect Misha Black and the City Planning Officer. The basis of the scheme was that the frontages to Magdalen Street (mostly local traders, plus some chain stores, banks, etc.) each contributed to a central enhancement fund, according to the extent of their frontage or their facade area. In return, their building front was renovated as necessary and redecorated as part of an overall street enhancement scheme. It was essentially a visual improvement exercise, to an attractively varied but co-ordinated artistic concept (using colour codes that were both in tune with Norfolk/East Anglian vernacular, and with modern 'approved' architectural paint ranges). Today we would tend to think of this approach as only 'cosmetic' but it fulfilled a useful purpose at the time in brightening up and making more attractive to shoppers in particular a lengthy shopping street on a major approach route into the city centre. Magdalen Street is not really an integral part of the main shopping 'core' of Norwich, but the scheme was important in helping to redress the balance rather than writing it off as not worthwhile. Though perhaps of limited effect locally, seen against the longer timescale, the Magdalen Street 'facelift' scheme nevertheless carried an important message for the future: *that visual enhancement of public 'activity' streets is well worth doing, as a means of both attracting economic activity and prolonging the life of traditional or historic streets which might otherwise fall more readily into progressive decay or decline.* It is probably a reminder also that in times of general economic recession, all kinds of extra measures are necessary to safeguard existing urban 'heritage', of which visual enhancement is an available option.

The private sector has to some extent turned its attention in more recent times to the need for enhancement of shopping or residential streets in the more well-established urban areas, and certainly meritorious instances exist of well-known firms of professional UK property advisers such as Donaldsons or Hillier Parker either co-ordinating projects or simply advising local authorities and business communities upon how these areas can be effectively improved, via street schemes where a broader approach is adopted to the need for general enhancement.

Grant aiding

On a wider front, however, the problems of increasingly decayed inner-city areas present an even greater headache in both planning and political terms than the problems of deteriorating historic streets or shopping areas such as Magdalen Street. Where the huge and often intractable problems of the inner cities are concerned, an extra element in the form of grant aiding is usually found to be essential for any real improvement to be made (and for acutely political situations to be assuaged). In 1994 the Single Regeneration Budget (SRB) brought together the government's existing expenditure on its twenty separate inner-city programmes in England, some £1.4 billion in 1994/95. The budget has the aim of providing flexible support for urban regeneration. Almost £1 billion is accounted for by land, property and development-related regeneration.

The Urban Regeneration Agency, later renamed English Partnerships, was established in 1993 to bring vacant, derelict or contaminated land back into use for housing development, commercial and industrial development, recreation or open space. It has taken over City Grant, Derelict Land Grant and the work of English Estates in providing commercial and industrial space, mainly in 'assisted areas'. It can compulsorily purchase land and designate areas for longer-term strategies to tackle dereliction, including controlling development and providing road links. The aim is to work by agreement with local authorities in identifying sites and reclaiming land, and with the private sector in developing the land.

Although these measures are likely to achieve some greater flexibility, resources are allocated to areas and projects on a competitive basis and the overall level of financial support from government for urban regeneration has fallen in recent years.

The SRB was prefigured by City Challenge which began in 1992. That mechanism combined the resources of a number of government departments with those of local authorities and private and independent sectors to restore confidence, initiative, enterprise and choice in the inner cities within selected areas of the country. Grants were paid by the DoE to selected local authorities to assist private sector projects which stimulated the local economy, provided employment, revitalised redundant buildings, and improved environmental and social conditions within their inner-city areas.

Under City Challenge, the basis for DoE grant aiding shifted from being related purely to inner-city areas with high social deprivation to a wider, more competitive context for claiming grants, e.g. where financial aid was seen as best directed towards reviving/improving economic prosperity in strategic areas or locations, i.e. where a greater chance of commercial/economic return existed.

The range of financial support for regeneration and development projects is very varied but the following is a guide:-

Department of the Environment

Information on the range of assistance administered through the SRB and similar schemes in Scotland and Northern Ireland is available from the following addresses:-

England	Department of the Environment, 2 Marsham Street, London SW1P 3EB
Scotland	Industry Department for Scotland, Urban Renewal Unit, New St Andrews House, Edinburgh EH1 3TS
Wales	Urban Affairs (ERP4) Division, Welsh Office, Cathays Park, Cardiff CF1 3NQ
Northern Ireland	Department of Environment for Northern Ireland, Clarendon House, Adelaide Street, Belfast).

Other grant-aid sources

Civic Trust (17 Carlton House Terrace, London, SW1Y 5AW). Under the auspices of the Civic Trust, a partnership was set up between Shell UK Limited and five national conservation organisations, to help local voluntary and community groups carry out practical environmental and conservation projects which benefit the community. As well as information and advice, small grants of up to £500 are available to help in the purchase of equipment and materials for projects undertaken by volunteers, whether in rural or inner-city areas. (Initiative administered under the Shell Better Britain Campaign, Red House, Hill Lane, Great Barr, Birmingham, B43 6LZ).

Royal Institute of British Architects (66 Portland Place, London, W1N 4AD). The RIBA has administered in recent years a Community Projects Fund from the Department of the Environment and Development Commission, designed to assist community groups in their search for capital funding through grants awarded to them, to cover half the cost of feasibility studies (up to a specified maximum), for community building projects or environmental improvement studies.

English Heritage – the Historic Buildings and Monuments Commission for England (Fortress House, 23 Savile Row, London, W1X 2HE).

- **Conservation Area Grants**[39] Two types of grant are available, for preserving and improving the character and appearance of conservation areas, as well as for repairing historic buildings and for associated environmental work, as follows:-

  (a) Town Scheme Grants applied for via local authorities who then apply to English Heritage for funding to cover normally up to 40% of the eligible costs of the project. Basis generally that of the conservation area overall, or the historic area forming the context of buildings for repair.

  (b) Section 10 Grants available directly from English Heritage, usually up to 25% of the costs of the project in question.

- **London Grants** English Heritage also awards grants to any worthwhile project within Greater London which affects historic places or listed buildings. Level of grant depends upon the type of project, its location, other funding secured, etc.

National Heritage Memorial Fund (10 St James's Street, London, SW1A 2EF). This fund provides assistance towards the cost of acquiring, maintaining or preserving land, buildings, works of art, and other items of outstanding interest which are also of importance to national heritage. A mixture of government grants and income received in investment makes up the financial package. The fund is designed as a safety net, i.e. where funding is essential to prevent loss or irrevocable damage in any particular case.

Grants and loans from the EU – via The Commission of the European Community (London Office, 8 Storey's Gate, London, SW1P 3AT). There are various funds and departments within the EU which finance urban initiatives, though guidelines and application methods are complex and change frequently. The EU has an overall policy of giving priority to less-developed regions of the Community, to increase cohesion by narrowing the gap between rich and poor regions. In the UK, the DTI designates Assisted Areas which have priority for EU funds. Typical target areas at January 1989, for example, were urban communities, areas of industrial decline and the long-term unemployed.

Pedestrianisation areas/schemes

Probably the most distinctive, and certainly the most public feature of urban change in post-war planning has been the emergence of the pedestrian area, in its different forms. The idea is not new, dating back to the Greek market area and the Roman forum, and beyond. Pedestrian areas, mainly in the form of the village, town or city square, have been created at almost any point in history either purpose-designed and built from scratch, or adapted from the existing urban fabric by the full or partial exclusion of wheeled traffic. Examples of both types of pedestrianisation of all shapes and sizes are plentiful in history, and are usually to be found where related originally to a primary function, as in the traditional European marketplace, the cathedral precinct, the square in front of the city hall/*palais de justice* or the formal approach to the seigneurial mansion/palace. Some of these, usually those planned in the 'grand manner', continue to reflect in a clear and undiluted way the single original purpose of their layout (such as in the Place Stanislas, Nancy, or St Peter's in Rome). The 18th-century London squares and the green spaces embraced by the crescents and terraces of Bath, Brighton, Cheltenham and Edinburgh show how successfully Britain has adopted the principle of the large or formalised pedestrian area for its own distinctive residential purposes. Continuing modern examples of the enclosed urban space in the 'grand manner' are still being built by the French, e.g. at La Défense in Paris, or even more recently the progression of urban spaces built on a grand scale at Montpellier. The English tradition, however, when not overtly influenced by the Continent (and by France in particular) is towards smaller and more informally planned spaces.[40]

Our tradition is also now much more towards the varied or mixed-use assemblage of buildings such as in the many marketplaces of different ages which are to be found throughout Britain, and particularly in England (the small Norfolk town of King's Lynn has two such spaces). Many of these central spaces in our towns still provide a main focus for urban activity, e.g. for shopping and offices of a 'local' scale, though major concentrations of these commercial activities tend nowadays to be built in their own separate central precincts (such as the very successful enclosed Ashley Centre in Epsom or Lion Walk in Colchester, or Churchill Square in Brighton, with the latter now in 1996 undergoing extensive redevelopment to a new layout). Other shopping precinct developments, of even larger size, have been built on 'parkland' sites at the edge of urban areas, like the American 'out-town' shopping malls, though the wisdom of this in terms of effect on 'downtown' trading is highly questionable. The main problem of city centres, however, is that of accommodating the access and parking needs of the car-borne modern shopper. Office precinct developments of a new, separate, purpose-designed nature also continue to be built, either redeveloped from the existing urban fabric, such as at Broadgate (the former Broad Street Station site in the City of London of highly rated modern concept and design, by Arup Associates) or at Canary Wharf in London's Docklands, where a lavishly-built modern office scheme lay severely under-occupied after completion due to the 1990s recession, plus over-optimistic forecasts of floorspace 'take-up' and of investment potential.

Apart from such specialised and purpose-built new developments as these, the major opportunity in Britain (and elsewhere in Europe) will be the adaptation of existing urban streets and spaces for primarily pedestrian use, enhancing at the same time the use (and prosperity) of the buildings lining the traffic-free spaces thus created – whether in the form of full or partial pedestrian areas or zones.

In Britain, the first of these central area pedestrian streets, or 'foot streets', was probably Butcher Row in Salisbury, Wiltshire, though it was not long before it was followed by London Street in Norwich, described earlier in this book and widely publicised. Both London Street and the term 'foot street' were the concept of Alfred Wood, the City Planning Officer of Norwich in the 1960s. 'Alfie' Wood (as he was widely known) had studied major Continental examples of pedestrianisation in Düsseldorf, Essen,

94 PLANNING FOR URBAN QUALITY

Figure 14 London, Broadgate/Norwich, Castle Mall: Award winners.
Both schemes won the RTPI Silver Jubilee Cup, presented each year to developments showing an exceptional level of planning achievement – the Broadgate scheme (above) in 1992, and the Castle Mall Shopping Centre scheme (right) in 1994.

In the judges' view, Broadgate (adjoining the newly rehabilitated Liverpool Street Station in the City of London) 'represents a model for a high-density city centre redevelopment . . . which has resulted in an unusually user-friendly, well functioning, mixed-use city centre environment with optimum public transport accessibility. In the case of Norwich, a major shopping mall was successfully integrated into the Castle hilltop site by building it underground on four levels with toplighting provided as part of a landscaped urban garden.
Sources: London, Broadgate: RTPI Journal, 11.12.1992; Norwich, Castle Mall: RTPI Journal, 2.2.1995

TOWN PLANNING FRAMEWORK 95

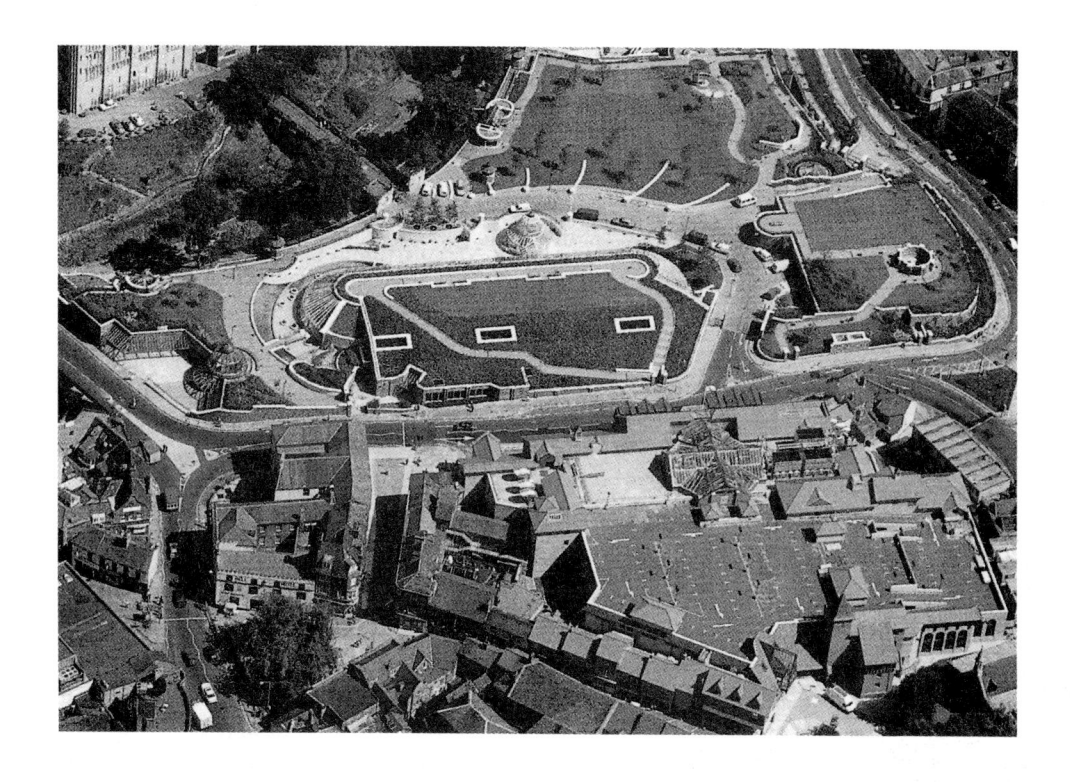

Cologne and Copenhagen, prior to implementing his scheme for London Street, his researches then concurrently published in his report *Foot Streets in Four Cities*. His initiatives in the area of street pedestrianisation in historic towns and cities in particular have become model terms of reference for others following in his footsteps.

But pedestrianisation cannot be achieved without the necessary redesigning, or at least intelligent adaptation, of the road system, to continue to provide vehicular access, parking and foot access to the central urban areas in question. Here again, Norwich proved to be the 'brand leader' in Britain, through Alfred Wood's ring and loop traffic system for the city, whereby internal loop roads provided access from the ring road system into the Norwich Central Area.

Other traffic arrangements may equally enable the creation of pedestrianised areas or streets, according to circumstances. In Epsom, Surrey, for example, the County Highways Authority scheme for improving the flow of central area traffic via a one-way gyratory system around the Ashley Centre enabled partial pedestrianisation and a supply of short-stay and disabled parking in a substantial area of the old high street, where it widened out in a curved frontage alignment for historical reasons. This satisfied the concerns of both the borough council and its Planning Officers, and the County Engineer, plus the public.

In other places, the need has been to create completely new traffic-free areas out of derelict or damaged urban environments. As far back as the 1950s, there was a desperate requirement in heavily war-damaged Coventry Central Area for a pedestrian shopping area. The cruciform-shaped central precinct achieved by Sir Donald Gibson did just that, and again was a forerunner of its type. Other similar shopping precincts were also built in the World War II bombed cities of Bristol, Exeter and Plymouth – though they, like other precincts of their time, have had to be updated and improved to answer later, more sophisticated environmental standards, plus different retailing patterns and size and design of shops and stores. In Rotterdam, Holland, the initially sparkling new 1960s Lijnbaan shopping precinct, built to help restore and renovate the bombed-out heart of the city, had by the 1990s sadly deteriorated to a great extent owing to vandalism - apparently largely unchecked in Holland – and possibly owing also to a rather dated form of linear layout (as the name of Lijnbaan implies) with only side and cross-canopies to guard against inclement weather. It is in this respect that enclosed centres, like the Ashley Centre in Epsom, score by comparison, though this should not preclude some sophisticated open-air forms of shopping precinct being developed. Epsom, however, rejected a similarly excellent open scheme designed in true conservation area vernacular style by one developer (Taylor Woodrow) in favour of an equally meritorious but enclosed scheme by another developer (Bredero).[42]

There is no doubt, however, that pedestrianised streets or precincts (to whatever degree enclosed) offer a greatly enhanced environment owing to the extra element of personal safety, convenience and relaxation offered to the visitor on foot, and that these traffic-free areas will continue to be provided as local authorities and private developers together realise their potential advantages. It is additionally significant that many old existing arcades and halls, like Covent Garden in London, frequently acquire a new lease of life for shopping and other public purposes, similar to the old covered markets at Bristol and Oxford, for example. Old railway station halls are often ideal in this connection: for example, Green Park Station in Bath, as a supermarket; part of Temple Meads Station, Bristol, for exhibition purposes; public art gallery, exhibition and museum uses now provided in the old Gare d'Orsay in Paris; the Railway Museum at York; and St Enoch's shopping centre in Glasgow. Most of these have

TOWN PLANNING FRAMEWORK   97

Figure 15 Traffic segregation: Bristol shopping arcades. Clifton
For the pedestrian shopper, freedom from traffic confers numerous advantages. There are also distinct benefits for established/historic town centres already suffering from competition by far too many out-of-town shopping malls. These examples from Bristol show how our older Central Areas can at least play an individual/complementary role to the out-of-town malls via sheltered traffic-free speciality shopping networks, full of ambience/character. Indoor markets and shopping malls were not always the product of the 'out-of-town' American dream inserted into any car-borne Western society. Bristol continues to enjoy a fresh food market in its old fish market, and a 19th-century arcade has been reopened in the centre of Clifton. A car isn't needed to take the shopper to either.
Source: Gordon Power

their own open forecourt areas in addition. (Here, we are at the meeting-point or overlap between the enclosed activity space and the open pedestrian precinct.) The opportunities are almost endless, given the amount of change that will inevitably continue to the use, scale and movement patterns in our urban areas generally and to individual buildings.

Detailed requirements and opportunities will arise at the same time for manifold items of 'floorscape' paving, street furniture (seats, lighting, bollards, bins, signing) apart from the overall need for 'urban landscaping'. By this latter term, as we have noted earlier, we do not simply mean 'greening' the urban scene by means of 'soft' planting (trees, shrubs, borders, climbing plants, etc.) though this in itself can be a very useful element in humanising and civilising a built environment. Urban landscaping is a term embracing the various means of shaping the urban scene, including designing for the rise and fall, opening and narrowing, light and shade of the environment, via the 'floorscape' in concert with the disposition of old and new buildings, and the single or multiple spaces they jointly create. The prime example in Europe of 'urban landscape' without any planting other than shrubs in boxes or jardinières is the Piazza San Marco in Venice, where the only additional features within the joint two spaces of the Piazza and the Piazzetta are the campanile, the line of flagpoles in front of St Mark's, the two columns fronting the lagoon, and the people sitting at the cafe tables or promenading. The superbly proportioned spaces do the rest. However, in most urban precincts – especially in Britain, with its very variable climate and atmosphere – 'greening the scene' is often an indispensable element.

Traffic restrictions and traffic-calming

Pedestrianisation is, of course, an extension of the principle of maximum traffic restraint, to which we have referred earlier. The latter is just as importantly applied at the local or 'micro' scale, i.e. to improving conditions for people living in ordinary residential streets, where their quality of life, safety, convenience and urban quality generally are reduced or threatened by traffic. There are a number of ways in which this can be remedied, apart from the obvious solution of directing extraneous traffic elsewhere (so far as this is possible without detriment to those other areas). There are two phases or stages in which this can be carried out:-

1 the experimental phase; and
2 the permanent stage.

In particular cases, road narrowing and the introduction of bends, curves, chicanes and the like may be the best solution (i.e. 'horizontal' types of measure),[43] together with speed restrictions and the creation of 'traffic exclusion' areas or zones, of whatever size or extent. The other type of 'traffic-calming' initiatives ('vertical'

measures)[44] include items such as speed humps or 'sleeping policemen' (not permissible on classified roads, however), or raised 'speed tables' of broader configuration. All of these are governed by detailed Highways Authority/DTp regulations in Britain, but the essential point is that where traffic exclusion or restraint can be legitimately and sensitively applied, the quality of life and of the environment both improve. A further point is that the detailed type of traffic-calming measures used should be suited to the nature and configuration of the environment in question, and therefore should be carefully chosen. Continental experience and practice may well be of benefit to the UK here.

IN PURSUIT OF URBAN QUALITY: PUBLICITY, CONTROVERSY AND EDUCATION

Introduction

A great deal is talked and written about quality in the urban environment today, and it is a hopeful sign that this should be so, despite (and possibly also because of) the lack of political will and resources to that end. *But quality means different things to different people, and no single aspect or interpretation is definitive – they all have their place.*

- The town planner, particularly in the public sector, will be concerned with 'achieving quality' mainly in the sense of improving the effectiveness of the town planning service. The outcome of that service, in terms of what it means for the built environment, should be of integral concern.
- The urban designer and the architect will both be largely concerned with the built environment in design terms, i.e. how the use of design skills can enhance or improve the quality of the urban scene. They will take both the broad 'contextual' view of the inher-

ent qualities of the area and the 'focal' view of how individual buildings should be designed/built/or treated within that overall context.

- The preservationist will judge an area by how well its historic buildings are taken care of, used and integrated with the urban scene generally as core features.
- The building contractor and the Building Control Officer will think of the urban fabric in terms of constructional standards and fitness for purpose, building by building.
- The chartered (commercial) surveyor will take a wide-ranging and hopefully balanced view embracing mainly the town planning, development value and commercial/marketing aspects of urban quality.
- The developer will be preoccupied chiefly with the financial risk element of his investment in sites and buildings, their capital value, and their monetary return, but will also need to bear in mind the wisdom and longer-term value of design quality investment in his development.
- The sociologist (like most members of the public) will primarily consider whether an area enjoys a strong sense of community as well as material benefits, e.g. good schooling, shopping, public transport and 'environmental quality'.
- Finally, and not least, we have the public at large, whether in the form of articulate individuals, or in the shape of amenity and civic societies, or residents' associations. This last category, it should be noted, now plays an increasingly active role in the planning and development world (and therefore hopefully in the process of achieving true urban quality).

It is as well, therefore, that we turn our attention to some of those very public aspects of 'the pursuit of urban quality': publicity, controversy and environmental education.

TOWN PLANNING FRAMEWORK

Influence of the Press and the Media

It is probably more relevant than ever to state that architecture, development and town planning are never very far from the public sphere of interest. This heightened public awareness is basically healthy, being the reflection of a concerned and involved society, but at the same time it can often become a vehicle for the expression of non-progressive and conservative British attitudes. An informed balance is always necessary between those seeking change and those resisting change. (A society unwilling to contemplate change is a society that is not even standing still but in fact going backwards).

Publicity, controversy and debate all have their place in the increasing public awareness of the process and influences that affect the outcome on urban quality. No longer are matters of architecture and town planning the sole prerogative of architects and planners – the combined interest and attention of the public and press have played a major part in effecting this change (however much the professionals have tried to erect barriers and retreat behind them).

Politicians likewise are not exempt from the debate. Even the financiers – those people who heavily influence and often determine the outcome of most kinds of development matters – find their lives increasingly affected by a combination of the planning system and, significantly, public attitudes towards planning, architecture and development generally – focused as always by a vigilant and influential press. For the debate on planning and architecture is also in fact an interface between conflicting strands in our life and society, often bringing into play deep-seated attitudes and prejudices.

The role of the press in setting out the public debate on controversial planning and design issues, and also in acting as a form of promulgator and popular educator in such matters, should not be underestimated. The press is often the only immediately effective means by which the professional architects and planners can get their message across to the public and, on the other hand, it is the main channel by which the public can voice its incomprehension or discontent as to what is happening to its environment.

Publications, Publicity, and Pronouncements

In terms of publications and publicity, a wide array of books, magazines and newspaper articles, radio and television coverage is available nowadays on matters of planning and architecture in Britain. There is little room for the public to complain of a lack of information, especially when press survey of popular or contentious matters regularly forms an adjunct to the public consultation mechanism of the planning system. It is right and healthy, in principle, that this should be so. Not only should the so-called hegemony of architects, planners, developers, financiers and local and central government be more open to challenge and debate, but it is very meet, right, and proper that the public should *be made aware of the issues and implications; and just as importantly, understand the criteria.* The national daily press in fact plays an important role in bringing to the public's attention important topics in the world of development and planning legislation, including central government's response to issues of major significance, which otherwise might not enjoy a particularly high profile in the public's perception.

Probably the most significant book on the role of urban design in the public realm to be published in recent times is Francis Tibbalds' *Making People-Friendly Towns*. The author (who died prematurely and tragically in 1992) was an architect-planner of international renown, and his book is a worthy and valuable successor to Gordon Cullen's well-known book *Townscape*, which was published in 1961, and still stands as a seminal work for architects and planners today.

While Tibbalds' book is a natural successor to Cullen's *Townscape* (even though the two are 30 years apart), it stands interestingly in contrast to another recent work, namely the Prince of Wales's 1989 book *A Vision of Britain* (and his TV film of the same title). Both Cullen and Tibbalds present the viewpoint of professional urban designers; Prince Charles presents both his own and (through his advisers) the layman's view of British architecture and planning. Whilst, as one might expect, both Cullen and Tibbalds as architect/planners take an outgoing if divergent approach to the question of urban quality based on a professional design-oriented view (Tibbalds is more pragmatic in planning terms), Prince Charles tends on the whole to take a layman's view of modern buildings and development, in terms of reaction against either their incongruity or their mediocrity.

The images and predilections which the Prince invokes in search of principles of good design or acceptable development are essentially rooted in the past, resting almost invariably on the concept of continuity and tradition. His concerns with our increasingly unsympathetic urban surroundings are however fully understandable. One of the Prince's favourite targets in his book (and elsewhere) is that of modern architects – another is modern architecture.

Prince Charles's well-known views on modern development were to some extent rebutted by another contemporary foray in this field, viz Maxwell Hutchinson's 1989 book *The Prince of Wales: Right or Wrong? – An Architect Replies*. This counterblast to royal predilection breezily exposes and examines several of the most cogent and topical issues of our times, and debunks the various popular prejudices, attempting at the same time to cast the light of professional knowledge and experience into areas still largely cloaked in misinformation and inaccuracy.

Major Architectural Controversies

Probably the three most famous planning and architectural controversies (i.e. the most significant) upon which the Prince of Wales not only made his views clearly known, but thereby also exerted a considerable influence on the outcome (tipping the scales in fact) were:-

- the Mansion House Square development, on the Mappin and Webb site, in the City of London;
- the National Gallery extension, in Trafalgar Square; and
- the Paternoster development in the City.

The protracted and painful saga of the extension to the National Gallery is now resolved, in as much as the classico-modern design (functional, but with Classical references and derivations) by the American architect Robert Venturi was finally selected and built. The turning-point in the saga was undoubtedly the intervention by Prince Charles, in his Hampton Court speech of 1984 at the 150th anniversary of the Royal Institute of British Architects. The two main 'items on the menu' at this glittering occasion were not only the acutely indigestible issue of Trafalgar Square/National Gallery but also the RIBA itself. The Prince's speech came just one week before the Secretary of State for the Environment (Patrick Jenkin) received the public inquiry inspector's report, and was clearly a major factor in persuading the trustees of the Gallery to jettison the uncompromisingly modern extension designed by Ahrends, Burton and Koralek.

The National Gallery competition had begun in 1981, with advisers appointed by the government via the PSA (Property Services Agency), but remote from the client (the Board of Trustees) and with a brief which called for a mix of gallery and office floorspace (34,000 sq ft (3,159 sq m) and 66,000 sq ft (6,131 sq m) respectively). With a brief that attempted to combine prime commercial office space with galleries of Renaissance paintings, and a pluralistic climate of various shades of modern architectural thought deployed upon the competition, it was perhaps unsurprising that the outcome should be a tangled and twisted tale of uncertainty, delay, reconsideration of the brief, appointment of architects ABK on the basis of need for a redesign, more delay, etc. Finally the Gordian knot was cut by the Prince himself, from his viewpoint that none of the schemes submitted had adequately respected (or even considered) what he felt to be the overriding context of Trafalgar Square and of the National Gallery itself. There are many who feel that:-

1. the saga should never have taken place; and
2. the finally executed design was, in the circumstances, the only possible outcome (certainly in the British tradition of compromise) if the architecture of William Wilkins' original National Gallery and the character of Trafalgar Square were both to be adequately recognised and reflected.

But the salient factor to emerge was, of course, the decisive intervention by the Prince leading to the solution (with the magnanimous financial support of Lord Sainsbury) which we see today, and which the public largely now endorses despite the bewilderment they must have certainly felt previously.

An even more protracted saga has arisen – one that is by no means yet fully resolved – in which the Prince of Wales has again played a pivotal role, namely that of the Mansion House (Mappin and Webb) site in the City of London. This was the issue originally of the glass and steel tower block designed for the site by the German-American architect Mies van der Rohe in totally uncompromising style, which was just as fiercely opposed. As is well known, it was referred to by Prince Charles in his Hampton Court speech of 1984 as a 'giant glass stump' (he had referred to the ABK design for the National Gallery extension as 'a kind of vast municipal fire station, complete with the tower that contains the siren', and again as 'a monstrous carbuncle on the face of a much-loved and elegant friend'). As with the National Gallery extension, in just one phrase the Prince had dealt a mortal blow to Peter Palumbo's plan to develop the Mappin and Webb site, facing the Mansion House, with a scheme and a design that Palumbo felt was in tune with the times (if not demonstrably the architecture of the Square). The saga of the Mansion House site nevertheless continued. Mies' towering steel and glass design (which had made no concessions to its central historic context) having been killed off, Palumbo then switched his patronage to an equally eminent modern architect, James Stirling. This represented a significant change away from the American-German 'big business' ethic of the highly functional and modern, but uncompromisingly monumental tower, dominating (indeed dwarfing) the important historic buildings across the intended new plaza (the Bank of England and the Mansion House), towards what Stirling intended to be a more human-scale solution, containing subtle responses to the site's urban context (via changes of plane, projecting corners, roofline variations, articulations of mass, etc. – if indeed a coherent design embodying so many subtleties could thereby emerge).

Unfortunately, Stirling had acquired something of a chequered reputation through his own prickly 'stand-alone' character and his track record of earlier buildings in Britain which, whilst being bold and innovative, also unluckily developed problems of either design or maintenance (e.g. Leicester University Engineering Department, where vertical facing tiles kept falling off, or Cambridge University History Faculty/Library, where the roof leaked, and the glass surfaces over the reading room faced the sun and heated up the room intolerably at times). Inevitably, these faults were soon used as a stick with which to beat not only Stirling but the architectural profession as a whole. But many buildings which leak, etc. do not get anything like the same amount of publicity, so the explanation must also have to do with the public view of 'intransigent' modern design. (However, compared, say, to some designs by modern Spanish or Japanese architects, Stirling's buildings are positively restrained. His Stuttgart gallery of fine art, the Staatsgalerie, is probably his most imaginative and has won widespread acclaim, even if the assemblage of curved and inclined window/wall planes may strike us as quixotic.)

Stirling's design for the Mappin and Webb site (One Poultry) was refused planning permission in June 1987 by the City of London (by two votes). A public inquiry followed, and in June 1989 the Secretary of State for the Environment (Nicholas Ridley) granted permission. The decision was contested by several conservation groups (an influential critic of modern buildings, Gavin Stamp, called One Poultry 'a monument to vulgarity'), who lost their case in the High Court in December 1989, then won it in the Court of Appeal in March 1990. Palumbo then took the case to the House of Lords, where the scheme was finally given the go-ahead in November 1991. This, however, was not the end of the saga – more delay ensued over a Road Closure Order which was needed in order to replace with a pedestrian way the narrow Bucklersbury Street running through the site, and the opponents of the scheme seized on this as another weapon.

Sir James Stirling died in 1992, but with the road closure problem then resolved (via a cash payment!) and construction of the scheme scheduled to begin in 1994, and with completion in 1997, Palumbo continued to back the scheme, despite the early 1990s recession, and despite the enormous gestation period involved.

The building of Stirling's monumental but bluntly modern design for One Poultry, in the heart of the City, will owe nothing to Prince Charles's intervention, and his apparent taste for historical 'repro'. But whether history will judge it to be worthy of the neighbouring grand classical buildings by Sir John Soane, Charles Dance or Sir Edwin Lutyens, whether it may even be judged a masterpiece (unlikely as that seems), or whether history and public opinion will reject it as unworthy, remains to be seen. (Certainly if it had to depend entirely on the somewhat conservative architectural views of the Prince of Wales and his many adherents, there would have been no chance of even 'getting it off the ground'). *The battle for the Mansion House Square site has embodied exactly,* as few other sites have, what one writer at the time described as *'the titanic battle between architectural modernism and traditionalism; between advocates of development and conservation'.* As in the Trafalgar Square saga, the struggle between these opposing forces has highlighted how the counterweight of public opinion on modern design in Britain, focused through a royal personage, and magnified through the press and the media generally, can have a terminal effect upon progressive architectural thinking and design. (The architectural world has had its own, equally acute, dichotomy; the debate over the main issue of the late 20th century, in design terms – that of 'Post-modernism'.)⁴⁵

The split between tradition and modernism goes right down to the very roots of our essentially traditional society. There is sadly little or no climate of intellectual enlightenment in Britain for the acceptance of avant garde (or simply new) architectural thinking on the part of the general public, and little or no climate of state or regal patronage. In France, successive Presidents have promoted, right up to today, intellectual or monumental buildings. In Britain, the heir to the throne appears to have initially sided with popular predilection against 'progressive' design of buildings, though latterly (by the mid-1990s) there has been evidence of a rather more accommodating approach on his part to modern architecture, via a greater emphasis towards the need for compatibility between old and new buildings. This is welcome, even if it does not quite add up to patronage of great works of public art or architecture as in France – in Britain, architectural patronage tends to be of a less tangible variety on the whole.

New Buildings in Historic Contexts

One common thread appears again and again in these disastrous histories of struggle, delay, defeat, appeal, delay, recasting, compromise and mounting cost: the recurrent failure to consider the essential context of the sites' surrounding characteristics, in particular where there are undeniable restraints of architectural heritage or historical imperatives.

Seminars are often revealing; at the 1992 RIBA seminar, Gavin Stamp said

> Architects are much too concerned with trying to be original. Originality is a snare and a delusion. Very few architects are really creative and actually advance the art . . . what we need is for them to be competent in producing an architecture which is not actually offensive . . . in other words we need a high standard of mediocrity.

This however begs the questions of change and development in architectural design and technology.

Many people, in Britain would subscribe rather more to this type of view than the idea of liberating and encouraging intellectualism in design, as in France. *Mediocrity, in the beneficent sense of the term* ('of medium quality, neither good nor bad', rather than the less commendable sense of 'second rate') *is a concept which the British are able to welcome and embrace as basically safe and uncontroversial – unlikely to prove damaging.* In fact, the very low standard of mediocrity, as Gavin Stamp puts it, which was evident in the 1960s (despite Gordon Cullen's influence) did immense visual damage to our towns and cities. Public opinion was not then sufficiently aware or mobilised to identify the danger, and the conservation movement had hardly started. Contrast this with the 18th century, he said, when almost any builder or architect could put up a competent building which respected its surroundings. At that time, of course, they were sustained by a rigorous and comprehensive tradition, as both Georgian London and Dublin testify. Stamp's view, however, was that architects should not copy Georgian facades; the 18th century enjoyed a cultural totality achieved by universally held values. As he explained:

> We cannot hope for that today. What we can hope for in historic contexts is general respect for such factors as heights, street lines and materials, although no rules can ever be hard and fast. Fitting into sensitive contexts requires more thought and choice than the average architect is capable of. For we have no one authentic 'modern' style today, rather a great range of approaches. Copying your neighbours is not necessarily the answer. I have no more faith in the conservative opinions of planning officers than I have in the often "exhibitionistic" pretensions of so many contemporary architects.

ACHIEVING QUALITY IN URBAN DESIGN

The Royal Town Planning Institute's annual Summer School is essentially a valuable and constructive forum for the exchange and presentation of new thinking on planning topics of all types. It is not confined to professional planners; every year there is a Councillors' School of one week's duration, which is of particular value in discussing topical problems and getting messages across within the planning system. There is thus a high degree of 'grass-roots interface' amongst those who have to administer the system and deal with the proposals generated by developers and architects.

One of the most significant presentations given by any speaker in recent years on the quality of urban design was the September 1991 paper by (the late) Francis Tibbalds, the Royal Town Planning Institute's 1988 President. In his address, Tibbalds – an eminent and talented architect-planner himself – set out 'a simple but non-negotiable position' *that good design must be a fundamental objective for both the planning system and the development industry.* Although accepting that design was a subjective matter, he argued that it should be possible to arrive at a set of principles to which all the professions concerned with the built environment can subscribe, and incorporate these into central government legislation and local planning policy. Tibbalds listed a number of his *bêtes noires,* e.g. that 'design' is seen as something to be sprinkled on once everything else had been decided. He complained of 'acres of classical wallpaper now being rolled out over big development schemes', 'fake-believe theme parks', 'pastiche and fakery', etc.

Tibbalds emphasised that the process of urban design should recognise the complex and organic nature of urban areas, and that the approach to them must be genuine and not artificial, contrived or too selective. Towns and cities, he felt, are not made up just of landmark buildings and civic monuments, but of 'backcloth' architecture that can be manipulated to make a rich and interesting urban environment which can also accommodate a few set-piece gems deliberately located in the townscape.

He saw encouraging signs of progress in the debate over design: the Prince of Wales's 10 principles of good design, his own urban design 'decalogue' and the set of 14 principles for 'user-friendly' environments produced by colleagues within his consultancy. Francis Tibbalds vowed that he would not be satisfied until the issue had been properly tackled at government level and proper policy guidance on design drafted and issued to all local planning authorities. He reminded councillors that design is much more on the agenda than it used to be, and said that it was now up to the institutes, the schools (of planning and architecture), local authorities and private practices to 'give an almighty shove' in the direction of better design as it may be some time before there is a similar chance.

He summed up with four main suggestions for 'getting it right':

1 There should be a greater commitment to good design from central government, with the responsible environmental ministers taking a greater interest in design, particularly in the public realm.
2 He sought radical changes in the training of the professionals concerned with the design of the urban environment. Architects, planners, engineers, etc. should not be at each other's throats, but should be natural allies with shared concerns for the physical environment.
3 Urban design needs to be properly recognised within local planning authority structures. 'It is more than a tame architect giving, on a part-time basis, design observations on a never-ending pile of very mediocre planning applications. It is about caring for the physical quality of the area as a whole – looking after its past and designing its future. It is about making things happen.'
4 Although design skills are important, so is a sensitive approach to the after-care and management of places (a woefully neglected aspect, on which there is little input and even less control), an understanding of the economic and social dynamics of change, and the ability to seize opportunities as they are presented. The community and the elected representatives were as much if not more involved with this as the professionals. Not only the concept but also the details must be got right, if development is to survive the passage of time, and adequate provision must be made for future maintenance, he emphasised.

The Prince of Wales's 10 Principles of (Traditional) Design

The foremost critic of bad and/or modern architectural design, and accordingly an overt exponent of criteria by which matters of design and development should be judged in the 1990s was notably the Prince of Wales. In his 1989 book *A Vision of Britain,* Prince Charles put forward 10 personal principles by which this type of assessment should be made.

1 THE PLACE We must respect the land. It is our birthright and almost every inch of it is densely layered with our island history.

The landscape is the setting for all our architecture. In the United Kingdom we are blessed with an enormous variety of landscape. Such richness – from the hills of Derbyshire and the Yorkshire Dales to the level fens of Lincolnshire – is greatly to be treasured. Often the type of land is gentle and subtle and correspondingly delicate and easily damaged. Much has been damaged – the last war didn't help in this respect – but it can be restored if people are prepared to take the time and trouble.

New buildings can be intrusive or they can be designed and sited so that they fit in. It is seldom enough to disguise them by planting: the scale and nature of the new buildings are crucial. Rather than planning from a drawing board or plotting road routes on a map, we should feel the lie of the land and its contours and respect them as a result.

New buildings should not dominate the landscape but blend carefully with it. Often large buildings can be separated into elements which will humanise the scale, give a gentler skyline and enhance the picturesque quality of our landscape.

We must protect the land. We need, for all sorts of complex historical and psychological reasons, to keep a sense of wilderness. The green belts are a valuable contribution to the preservation (even if sometimes this is an illusion) of the countryside. If new buildings avoid 'sprawl' and are grouped together, more of the landscape can be preserved. There's no point, as I see it, in having green belts unless they are genuinely green. It would help, in a small island, to avoid the wirescape, the spread of sodium lighting, the rash of badly designed supermarkets and petrol stations and the fields full of parked cars. Farms and many Government buildings and structures such as army camps and power stations may be immune from planning laws, but they should not be immune from aesthetic considerations – they, *above all*, should respect the land.

2 HIERARCHY   There are two kinds of hierarchy which concern us here. One is the size of buildings in relation to their public importance. The other is the relative significance of the different elements which make up a building – so that we know, for instance, where the front door is!

Buildings should reflect these hierarchies, for architecture is like a language. You cannot construct pleasing sentences in English unless you have a thorough knowledge of the grammatical ground rules. If you abandon these basic principles of grammar the result is discordant and inharmonious. Good architecture should be like "good manners" and follow a recognised code. Civilised life is made more pleasurable by a shared understanding of simple rules of conduct.

A good building that understands the rules explains itself in its forms and spaces, tell us where to go and what to expect. It emphasises those parts that are public and important. Even in the smallest house there is a distinction between back and front doors, between living-room and attic windows. Only in recent large buildings have we lost this sense of hierarchy, so that it is hard to discover whether the block at the end of the street is a hotel, office or civic centre.

Public buildings ought to proclaim themselves with pride, as they have in the past. A Doric portico on a bank, an elaborate window-trimming on the meeting room of the town hall, a generous and lofty entrance hall in an hotel all serve to set the scene and uplift our spirits. Nowadays the dogma of modernism ensures a deadening uniformity.

Public façades are not the only issue: there is also the need for some buildings to be bigger than others. There is a recognisable hierarchy in towns and villages that may seem obvious. Churches, public buildings, halls and pubs all have their scale and special sites. In a way they emphasise our values as well as our social organisations.

3 SCALE   Man is the measure of all things. Buildings must relate first of all to human proportions and then respect the scale of the buildings around them. Each place has a characteristic scale and proportion: farmhouses in Nottinghamshire may be tall and thin and in Northumberland they may be low and squat. It is high, and out-of-scale buildings that are most damaging.

Almost all our towns have been spoiled by casually placed oversized buildings of little distiction, carrying no civic meaning. Each area needs a suitable and civilised height limitation. In London, for example, helpful bye-laws and building acts from the 1890s until the 1950s imposed height limits. These rules made for an orderly and elegant skyline. Above a continuous cornice line rose turrets, domes, spires and cupolas that we all can appreciate.

The image of the city that inspires these rules is of an enclosed and contained city. As these rules and bye-laws have gradually been replaced by elaborate discretionary planning rules, we have witnessed the rise of out-of-scale buildings in inappropriate places.

The redevelopment of towns and cities once respected plot sizes, existing street patterns, parks and squares. Today, far too often, developers are allowed to assemble several small sites, regurgitating them as gargantuan out-of-scale developments that look like Gulliver in Lilliput.

Sometimes a great public building may dominate a city, but it will be the sort of building that reflects our aspirations, like our great cathedrals. We raise to heaven that which is valuable to us: emblems of faith, enlightenment or government. But this vision must also be supported by small-scale buildings which reflect our intimate lives.

4 HARMONY   Harmony is the playing together of the parts. Each building that goes beside another has to be in tune with its neighbour. A straggling village street or a wide city avenue which may consist of buildings belonging to many different periods can look harmonious.

Nowadays there seems to be too much insensitive infilling of a particularly jarring nature in our towns and villages. Buildings boast too much and forget their neighbours. We have lost that desire to fit in which was once so natural to us, and gave us some of our loveliest streets and compatible groups of buildings.

TOWN PLANNING FRAMEWORK 107

Whatever happened to architects' and designers' humility?

Because of the scale of our country it is more necessary to respect our indigenous roots than to imitate transient international architectural fashions. Our older towns cannot easily absorb the more extreme examples of outlandish modern design. Perhaps we should give the best contemporary designs a sporting chance in our new towns?

Towns such as Cheltenham and Bath exemplify the virtues of architectural harmony, not only in their layout but in their organisation of the smaller architectural elements such as door cases, balconies, cornices and railings. It was second nature to the 18th-century architect to add one house to another with a sense of concord and unity.

Today buildings are designed from abstract principles and are thrust, in the name of 'new' architecture and 'modern' functional requirements, into the carefully scaled and painstakingly adjusted cities of the past. Their impact can be softened by an acceptance of the existing street rhythms and plot sizes. Buildings in a city such as Edinburgh are the individual brush strokes of a grand composition, because all the participants understand the basic rules and traditions. Harmony is the pleasing result.

5 ENCLOSURE One of the great pleasures of architecture is the feeling of well-designed enclosure. It is an elementary idea with a thousand variants and can be appreciated at every level of building from the individual room to the interior of St Paul's Cathedral, or from the grand paved public square to the walled garden.

The scale can be large or small, the materials ancient or modern, but cohesion, continuity and enclosure produce a kind of magic. The application of these ideas makes a place unique.

The secret of enclosed spaces is that they should have few entrances; if there are too many the sense of security disappears. If the space contains something to love such as a garden, a sculpture or a fountain, it is more likely to be cherished and not vandalised. A community spirit is born far more easily in a well-formed square or courtyard than in a random scattering of developers' plots. The squares, almshouses, universities and inns of court of our past that we love so well have always answered our needs. Their virtues are timeless, still providing privacy, beauty and a feeling of total safety.

Surely we can apply some of these lessons to the vast and neglected soulless deserts of post-war housing? Our new housing estates need not always be strewn clusters of separate houses set at jagged angles along windswept planners' routes. Examples exist all around us of the ideal homes that people have loved for ages: it is simply a matter of learning to imitate the best. Discriminating observation of the past must be the inspiration for the future. It is surely a worthwhile investment in many cases to resurrect this age-old principle of enclosure which creates a recognisable community of neighbours.

6 MATERIALS Britain is one of the most geologically complicated countries in the world, and as a result it is one of the most beautiful. Our rich variety of building materials is a source of constant pleasure and surprise, for our villages and towns were built from what came closest to hand: stone in Northamptonshire, timber in Herefordshire, cob in Devon, flint in the Sussex downs, brick in Nottinghamshire. Each town and each village has a different hue, a different feel, and fosters a fierce loyalty in those who belong there. We must retain this feeling; we must ensure that local character is not permanently eroded.

While the 19th century saw the beginning of that erosion, the late 20th has brought a bland and standardised uniformity to our building materials. We can no longer tell where we are. Concrete, plastic cladding, aluminium, machine-made bricks and reconstituted stone are shipped to every corner of Britain from centralised production lines. This has created an overall mediocrity: a kind of architectural soap opera.

To enable new buildings to look as though they belong, and thereby enhance the natural surroundings, each district should have a detailed inventory of its local building materials and the way in which they are used. This should become a bible for local planning authorities and should be held up as a model to developers and their architects.

Britain has to revive and nurture its rural and individual urban characteristics based upon local materials. Perhaps there is even a case for reopening some of our great stone quarries. We must also encourage our traditional craftsmen – our flint-knappers, our thatchers, our blacksmiths – and involve them in the building of our future. This will in time engender an economic revival which is not dependent on centralised industries but which is locally based.

7 DECORATION   There seems to be a growing feeling that modern functional buildings with no hint of decoration give neither pleasure nor delight. The training of the modern architect rarely encompasses the rules of ornament or the study of past examples of applied decoration. There is no longer a universal language of symbolism, and the gropings of some critics towards the imposition of 'meaning' on what they call post-modern architecture has been fairly unfruitful.

This apparent vacuum can in fact easily be filled. There is a latent national interest in decoration. You only have to look at the thriving DIY industry! Long-lasting traditions of ornament go back to our Celtic forebears, and a glance into any ancient parish church can reveal amazing decorative secrets.

Many people think that a revival of classicism can help. It is certainly a universal language but it is not one that can be applied easily unless it is thoroughly learned. It is not the simple pastiche that some critics claim it to be: learning the classical language of architecture does not mean that you only produce endless neo-Georgian-style houses. Classicism provides an incredibly rich inventory of infinite variety.

In Britain there has always been a parallel stream of the Gothic and the Arts and Crafts movements. These rich traditions rest in the hands of skilled craftsmen whose rare talents need constant nurturing to ensure their continuation. They are there, so why don't we use them more often?

We need to reinstate architecture as the mistress of the arts and the crafts. I would suggest that the consumers are ahead of the professionals here. They seem to feel, as I do, that living in a factory-made world is not enough. Beauty is made by the unique partnership of hand, brain and eye. The results should be part of all new architecture, helping it to enrich our spirits.

8 ART   While decoration is concerned with repetition and pattern, a work of art is unique. Why is it that contemporary artists play such a small part in the creation of our surroundings? Architects and artists used to work together naturally; today they are worlds apart. Look at so many of the great buildings of the past, where the architect needed the contribution of the artists to complete the splendour of his total vision. Imagine London's Banqueting House in Whitehall without Rubens' great ceiling; or the Sheldonian Theatre in Oxford without its sculpted emperors.

How dull, in comparison, are some of our recent national landmarks. No artists were commissioned to adorn the National Theatre, which does not even boast a statue of our most famous playwright. Where is the art on the plans for the British Library? Architects and artists should be betrothed at an early stage in any major public project. It is no use just standing a sculpture on a plinth outside a new building, almost as a guilty afterthought. Art should always be an organic and integral part of all great new buildings. Sculpture and painting play an essential role in conferring on public buildings their unique social and symbolic identity, which architecture alone cannot. Their pictorial iconography is an essential complement to the architecture. Much more art should be commissioned.

It will be hard for this unity to come about satisfactorily as things are at the moment. Artists and architects might as well be educated on different planets. The principles by which art and architecture are taught need to be revised. There should be common disciplines taught to all those engaged in the visual arts.[46] Life drawing and a study of nature is as essential for architects as it is for any artist. Remembering common roots can nourish both these great arts for their mutual benefit and our delight.

9 SIGNS & LIGHTS   Far too many of the marks of 20th-century progress take the form of ugly advertising and inappropriate street lighting, apparently designed only for the motor car. The car and commerce are both vital to the well-being of the country, but it is the junk they trail with them that we have to tackle.

Much of commerce and retailing seems to have a strange affection for banal logos, and the result is that the country is littered with a proliferation of corporate images. These images may give the companies concerned a clear identity, but they can be crude and damaging in our older towns and in the landscape. The decline of the elegant shopfront with good lettering and its replacement by standard plastic signs is much to be regretted. There is something demoralising about great companies making so little effort to respect the places where, after all, their customers live. We could learn from some of our European neighbours about controlling signs, particularly within historic towns.

Traffic signs and street lighting are aspects of the visual world that need to be kept under control. On the great national motorway network as well as on British Rail and the London Underground there has to be constant vigilance about visual standards, especially the quality of lettering. Good lettering must be taught and learned; its qualities are timeless and classical in the broadest sense.

Street lamps and lighting on the main traffic routes are often excessive and cast an alien sodium glow over large areas of the country.[47] In parts of Britain there is no longer any real darkness or twilight, only an orange glare seeming to indicate a fire. Our towns and villages should be beautifully lit at night. Safety is not a matter of light intensity but of the overall quality of the surroundings. Many great cities of the world have retained a magic quality at night due to incandescent lighting. We should bury as many wires as possible and remember that when it comes to lighting and signs the standard solution is never enough.

10 COMMUNITY People should be involved willingly from the beginning in the improvement of their own surroundings. You cannot force anyone to take in the planning process. Legislation tries to make it possible for people to share some of the complex processes of planning, but participation cannot be imposed: it has to start from the bottom up.

The right sort of surroundings can create a good community spirit. Too many areas of our towns and cities have suffered from the mentality of planners who zoned everything, keeping work and home miles apart and encouraging commuting. Good communities are usually small enough for people to get together to organise the things they want. This is possible in the inner cities as well as in smaller towns and villages.

Pride in your community can only be generated if you have some say in how it looks or how it is managed. The use of local builders, craftsmen and architects helps to achieve this.

There is a great need for more experiment in the management of community development. The professionals need to consult the users of their buildings more closely. The inhabitants have the local knowledge: they must not be despised. People are not there to be planned for; they are to be worked with. In the creation of new communities the problems may be more difficult, but there is always local knowledge and that is where a community starts.

It is time for more experiment in the way we plan, build and own our communities. For example, new initiatives are needed to try and find ways to ensure that our surroundings are not entirely sacrificed to the car.

In other areas sharing is a good way to start. The fine facilities of so many local authority schools and colleges, for example, could have a much wider community use. Churches could have an extra role as part of the healing process, working with doctors and hospitals. Schools and universities could, at certain times, be open to people of all ages.

All these things are part of a broader interpretation of 'planning' and 'community'. There must be one golden rule – we all need to be involved together – planning and architecture are much too important to be left to the professionals.

Although it is possible to find exceptions to, or to fault, some of these principles, they serve as a good starting-point. But what are they principles of, and by what standards? Essentially, they are intended as a reminder of the 'ground rules' by which the best of our historical environment and buildings were created. By projection, they also represent a profound wish to return to those criteria and standards for all future development. Apart from the last of the 10 principles, i.e. that of Community, they are all physical/historical in nature. This is not necessarily to decry the Prince's efforts to bring us back to a vision of our visually rich building inheritance, and of the complexity of our social and physical circumstances which created this richness of inheritance: in his own way, he has done us a service in reminding us of our roots, and in causing us to question the direction in which we are going. But if the result is merely to make us stop in our tracks and look longingly in a backward direction, ignoring the various new potentials that exist, then the message is to that extent misplaced and counter-productive. Principles of good design are timeless, but building forms must be flexible and reflect progress validly within context.

Prince Charles's principles for creating buildings and areas directly compatible with the past carry us part of the way towards a better urban environment, but only part way. *They do not, for instance, consider in primary or 'mainspring' terms the need for town management as a major catalyst of quality of life and of our surroundings.* Sensitive urban management is critical if we are to revive our decayed and unattractive cities by making the public realm agreeable for inhabitants, visitors, workers and shoppers, in which the car has to be accommodated as part of a package of measures involving traffic-calming, foot streets and public transport. Our traditional cities can provide an enhanced quality of life, with their mixed-use, non-traffic-dominated enjoyable spaces. In managing our cities, it is evident that we need to understand the 'grain' of the urban fabric and make the best use of what we have got, knowing what not to touch. This in turn provides an extra context for new design and development. The context must always be the basis!

Prince Charles in *A Vision of Britain* deplores the increasing complexity of the planning system, with its myriad rules and regulations, its network of legislation, and the general mill of bureaucracy, committees, negotiations and often long and expensive public inquiries. He would prefer a much more simple and straightforward system (in which good design would therefore flourish), based on an organic and 'natural' mixture of uses and buildings, held together by a common human scale.

The Prince was basically right in seeking a more clear and straightforward planning system, which enables good development to be created (whatever we may think about questions of architectural style). He would have done better, however, to concentrate more not only on urban management aspects, referred to above, but also on better relationships between developers and planners. More consideration has to be given to the role of developers, as the providers of new buildings, etc., as well as to the planners who control them. At its simplest level, the developer wants the planning system to protect his investment; the planner needs the developer to implement change (where change is necessary and for the general good). From the developer's point of view, urban design may be defined (as a leading UK planning consultant has) as 'the design and management of the space between buildings and the design of the building with references to its context'. It is not though a good basis for best results in urban design terms if either planners or developers are obliged to operate from a position of weakness relative to each other, either in operating the local government planning system, or in terms of enabling development to proceed on a financially sound and remunerative basis. If either side is hamstrung in this way, an unsatisfactory situation will soon be reached, and the desirable aspects of urban design quality are bound to suffer. Since a high-design quality is what both parties (and of course the public) should be seeking, a viable set of ground rules for achieving quality on all counts in development and planning matters is paramount.

Francis Tibbalds: Urban Design Guidelines/Personal Appreciation

The most succinct and appropriate set of guidelines for new architecture and the built environment to be produced in response to the Prince of Wales's Vision of Britain *TV film ultimatum*[48] *were without doubt those put forward in December 1988 by the late Francis Tibbalds (then President of the Royal Town Planning Institute).* Tibbalds' 10 commandments in urban quality were expressed very simply:

1. Thou shalt consider places before buildings.
2. Thou shalt have the humility to learn from the past and respect the context of buildings and sites.
3. Thou shalt encourage the mixing of uses in towns and cities.
4. Thou shalt design on a human scale.
5. Thou shalt encourage freedom to walk about.
6. Thou shalt cater for all sections of the community and consult with them.
7. Thou shalt build legible environments.
8. Thou shalt build to last and adapt.
9. Thou shalt avoid change on too great a scale at any one time, and
10. Thou shalt, with all the means available, promote intricacy, joy and visual delight in the built environment.

If Prince Charles has been a prominent publicist for the cause of traditionalism in architecture and planning, Francis Tibbalds has gone further in illuminating the whole subject of the quality of the urban environment and providing clear guidance on how the transcendent values of urban design and environmental quality should be approached and created. The RTPI Journal *The Planner* printed a tribute to him on 7 February 1992 in the words of colleagues who knew him well:

> The world of planning and urban design lost one of its most talented, charismatic and outspoken figures with the untimely death of Francis Tibbalds at the age of 50. For many if not most planners his name became synonymous with 'urban design', but he was very much a people's planner and a planner's planner. *Urban design was not seen by him as some kind of aesthetic magic dust to be sprinkled on to dismal environments. . . . It was a process of fusion of elements of planning, architecture, and the other relevant disciplines to one common purpose – the creation of user-friendly environments responsive to the needs and aspirations of real, ordinary people.* Central to that process in Francis's vision of urban design were public participation, the role of the planner as interpreter, facilitator and communicator, and the framework of the statutory planning system, warts and all. . . . Francis Tibbalds re-established urban design as a worthwhile, indeed essential, activity for the future development of our towns and cities.

Tibbalds always took the broad view, championing the causes of good urban design, conservation of urban heritage, and the improvement of the public realm in his work on major projects, including the King's Cross Development and the ground-breaking contribution of the Birmingham Urban Design Studies (see Part III, pp. 174–8). His 1990 report on Central Melbourne, in the form of a planning overview, was hailed as a seminal work, and shortly before his premature death in January 1992 he had been active on the Spitalfields redevelopment design panel with architect-planner Roy Worskett and Sir Philip Powell. He set up the Urban Design Group[49] in 1979, and remained its Chairman until 1986 while it grew in size and influence (it now operates in a European as well as a UK context). In 1988 he became President of the Royal Town Planning Institute, and his presidential theme 'Closing the Gaps' not only built bridges between planners, architects and the other environmental disciplines but also encompassed much wider social and political issues. He lobbied vociferously for political action on homelessness, investment in public transport, strategic planning, and a democratically accountable London-wide authority. He stood up for the career development and the rights of planners to stick to their professional views despite political pressures, and he gave wholehearted support to good planning education, lecturing widely all over the country and managing also to give talks to primary schools, careers conferences, and community groups and societies. In 1989 he was awarded the Faculty of Buildings Literary Award for his 'Ten Commandments of Good Urban Design', quoted above – his popular but controversial response to the public challenge made by Prince Charles.

The Prince may have raised his personal standard and rallied the forces of traditionalism in

Britain; Francis Tibbalds, however, led the crusade for a better future urban environment. His book *Making People-Friendly Towns* (illustrated with many of his drawings and published after his death in 1992) will no doubt stand as testimony to his widely embracing concept of urban design and as a main reference point in the creation of environments of both quality and relevance to the public's needs and aspirations. We quote from some of the more fundamental chapter summaries of that book, in the form of the Recommendations/Action Checklists set out by Tibbalds, and invite contrast of this approach with that of the 10 principles from Prince Charles's *Vision of Britain:*

Places matter most

1 The first priority is to agree what sort of public realm is appropriate in any particular area and then to agree the buildings, development and circulation system which are appropriate to it. Usually this is done the other way round, with devastating results for the urban fabric.
2 Places need to offer variety to their users. They need to be unique and different from one another — each rooted in their own particular historical, geographical, physical or cultural context.
3 In most instances, individual buildings will be subservient to the needs and character of the place as a whole. If every building screams for individual attention, the result is likely to be discordant chaos. A few buildings can, quite legitimately, be soloists, but the majority need simply to be sound, reliable members of the chorus.
4 Many town centres are small enough to be considered as single places. In the larger towns and the central areas of cities, over time, areas of different character are probably discernible. These should be defined and developed, providing they are for real, rather than artificial bits of make-believe or urban theatre that will, in the long run, devalue reality.
5 Try not to view the organisation or reorganisation of towns and cities purely from the rather exclusive points of view of the motorist or the developer. It is of greater importance to consider the needs and aspirations of people as a whole — with priority being given to pedestrians, children, and old people. This simple change or widening of priorities could, by itself, transform our urban environment and lifestyle.

What are the lessons from the past?

1 A clear and comprehensible framework and organisation must be devised, to which the various public and private agencies involved in implementation can relate.
2 A series of simple design rules and principles should be developed.
3 There must be strong and passionate commitment to quality, completion and maintenance from the town or city's leaders: no great city has been realised without the support of strong individuals.
4 Designs and plans should avoid factors which militate against achievability — such as placements of land-uses and communication routes (the 'traffic architecture' of the 1960s) in complex, single-purpose or inextricably close relationships to each other.
5 Old buildings will usually be devalued by copying, pastiche, or facadism (the practice of retaining only the facade of an old building and developing a completely new structure behind it).

The book continues with a number of chapters on specific aspects of planning and urban design, relevant to environmental quality. These are: Mixing Uses and Activities; Human Scale; Pedestrian Freedom; Access for All; Making It Clear; Lasting Environments; Controlling Change; and Joining It All Together.

Making People-Friendly Towns concludes with 'A Renaissance of the Public Realm', in which the 'action checklist' and final list of recommendations are as follows:

- To everyone:
  1 Look at every proposal again and again. How can it be made better? How does it square up to the preceding axioms?
  2 We must all care more about the physical environment and believe in good design.
  3 We need to foster a more open, collaborative approach amongst all participants in the development process.

4 We need to identify leaders who will look after their own towns and cities, encourage the right things to happen and stop the bad things.

- To central government:
  1 Give greater priority to the physical environment and the long-term future.
  2 Promulgate clear design advice in ministerial circulars and policy statements.

- To local planning authorities:
  1 Recognise the importance of urban design.
  2 Appoint appropriate personnel at all levels of seniority to handle urban design tasks.

- To the professional institutions:
  1 Break down professional demarcations. The environmental professions should all be natural allies, working together for the good of the environment.
  2 Encourage in particular the training of people with urban design skills.

- To the academic institutions
  1 Break down narrow, insular, teaching practices.
  2 Encourage multi-disciplinary working and studying at staff and student level.

This is followed finally by a Postscript, in the form of Model Guidelines for Design and Planning Control. These were drawn up from the text prepared by Francis Tibbalds in 1989 for discussion between the RTPI, the RIBA, the development industry and the Department of the Environment, UK. The text was formally submitted to the Secretary of State for the Environment in March 1990 as a suggested basis for a Ministerial Circular or Planning Policy Guidance Note on Design and Planning Control.[50] Such guidelines could form a useful basis, Tibbalds felt, for the drawing up of guidance which is specifically relevant to a particular town, city, region or country. They summarise the principal tenets of his book.

Other Influential Publications

We give due recognition again to the role of publicity, the media, and publications generally (whether books, magazines, professional journals, radio and/or television) *in promoting an awareness of the importance of good urban design amongst all those concerned either directly or indirectly with the built environment.* Publications such as Ian Nairn's *Outrage* and *Counter-attack* drew the attention of the public as long ago as the mid-1950s to the visual squalor of urban 'subtopia' – the mindless creeping blight of bricks, concrete and wirescape being allowed to spread without sensitivity, plan or even perception of the huge visual damage caused to existing areas of character and charm. Nairn also voiced acute concern over the creation of new areas of total non-identity through the brutalising effects of the lowest possible budgetary and creative standards applied to the urban fabric, in both social and physical terms – where aesthetics is a concept to be either derided or omitted altogether. Nairn's two books (illustrated by Gordon Cullen's drawings and copious revealing photographs) were originally published in the *Architectural Review* in 1955 and 1957 respectively and still stand today as the standard works of reference to the despoliation of our environment by a philistine society. It is a matter for judgement as to how far we have progressed since then. The planning system, so frequently criticised for encouraging the mediocre at the expense of the meritorious and the progressive, has at least kept the worst at bay over a wide front and a long period. Not all of the criticisms directed at it have been fair, but the charge of encouraging a general level of 'familiar mediocrity' cannot be entirely rebutted. (The way to hell is still paved with good intentions!)

Gordon Cullen's own book *Townscape*[51] soon followed up Ian Nairn's pioneering works during the 1960s and blazed a brilliant and shining trail forward, in civic/urban design terms, for the optimistic and heady days of the 1960s and the new generation of architects and town planners, to whom Cullen's approach to the urban environment (at once both traditional and modern) came as an invigorating breath of fresh air.

Now in the 1990s, 30 years on from those initially progressive days when there seemed to be all to play for, a very different atmosphere prevails. The national economic and political scenario in the 1980s was the market-led, profit-orientated aggrandisement of the Thatcher years, when private sector development pressures became 'ultra bullish' and resistance in the form of public sector control in the interests of general public amenity were officially discouraged by political dogma. This was swiftly followed by the world recession of the early 1990s, and the bullish tendencies of the development market gave place to some extent to a more conciliatory attitude towards the public, in whose domain or environment all kinds of new development was being built.

As had happened at fairly frequent intervals from the 1950s onwards (when the New Towns attempted to deal with post-war population pressures), social aspects of development in terms of the need both to cater for and to consult the public (not always synonymous) came to the fore again by the late 1980s/early 1990s. Whether this was more due to the persistence of enlightened public figures in politics, town planning, civic amenity movements and the universities, or whether due simply to the depressed state of the market creating room for a more socially oriented outlook, is again a matter for judgement. Journals such as *Town and Country Planning* show that the 1980s development boom was often being challenged by community groups and others, and indeed the House of Commons Employment Committee in their report 'The Employment Effects of Urban Development Corporations' (1988) was very critical of how the Docklands development excluded unemployed or unskilled people.

A generally related issue here is that of social displacement. It is essential, when considering redevelopment or traffic reduction/calming, that such measures do not in fact displace people or traffic elsewhere, i.e. when improvement is at the expense of others. (The issue is discussed in Alterman and Cars (eds) (1991), *Neighbourhood Regeneration: An International Evaluation*.)

Whilst the growing problems of world-scale pollution and deprivation focused attention upon the urban scene, the Green movement, via leading exponents such as Jonathon Porritt, were able to demonstrate and establish the virtues of 'sustainable development' (i.e. organising for maximum efficiency and minimal environmental impact, and reducing use of external resources) and point to the undoubted advantages, for example, of 'greening' the urban environment. Both these philosophies, which are complementary, receive increasing attention in the planning and environment press, and even make more headway now in political circles, though the invidious Thatcherite legacy of the market-driven political economy is far from dead, and still has the potential to frustrate (if not defeat) the best intentions of beneficent planning and thinking as affecting urban quality and design standards – and as affecting ordinary people.

As a reaction, however, to the oppressiveness of unguided market forces or to plain philistinism in the field of development, it was fortuitous (and perhaps inevitable) that the early 1990s saw the emergence of freely publicised principles of 'good practice', where both the protection and enhancement of the urban scene were concerned. We have referred to two of these already – Prince Charles's *Vision of Britain* and Francis Tibbalds' *Making People-Friendly Towns* – and this book is of the same lineage, though with the emphasis clearly upon the constructive and beneficial role of influence and cohesiveness which the public sector, in the form of local authorities, could and should play to a greater extent (given the necessary resources).

Regarding the growing public awareness and enlightment in some quarters, evident by the

1990s, it is significant that the popular design monthly *Perspectives*, which was given birth by Prince Charles's Institute of Architecture in April 1994, now reflects the need to reconcile the new with the old, and to design with people and users chiefly in mind, whilst pursuing quality and excellence of form.

Promotion/promulgation of standards for new development has not only been via publications such as these. The sporadic conflict between developers' architects and local authority planners over the vexed question of design control, and the concern of both the relevant professional institutes as to that situation, resulted first in the Design Accord between the RIBA and the RTPI followed by Annex B to the government's Planning Policy Guidance Note 1 (PPG1), both of which were circulated to local planning authorities only a year or so before the publication of Francis Tibbalds' book covering (*inter alia*) the same areas of concern. In fact, Tibbalds had put out his own set of model guidelines via a text prepared by him in 1989 for discussion between the RTPI, the RIBA the development industry and the DoE.[52]

In passing, the additional comment should perhaps be made that Frances Tibbalds and many others like him were at work on central issues of change long before Prince Charles entered the fray, and felt betrayed by his populist form of intervention.

Another publication worthy of particular note in our survey of prime contributory material is Judy Hillman's booklet *Planning for Beauty*, produced in April 1990 for the Royal Fine Art Commission. This also examined, in a highly succinct manner, the question of the relationship between aesthetics and planning control, as 'one of the major and most intractable architectural issues of the day', in the words of Lord St John of Fawsley, Chairman of the RFAC, in his introduction to the booklet. Judy Hillman considered this central question against a wide analytical backcloth, including aspects such as present practice, two-tier planning, current/recent developments, and potential areas for change. Her study, together with its recommendations, has been well received in all quarters and is warmly commended.

These publications, together with numerous articles in the professional planning and architectural press paved the way for further governmental guidance in 'Plans and the Environment' contained in PPG12 (Annex A), issued in February 1992. This document was hailed by Michael Welbank, the RTPI President, as 'the dawn of a new era'; not an overstatement when seen in the perspective of the 1980s and 1990s.

Design Control: the RIBA/RTPI Design Accord

Following the stagnatory regime of the 1980s, when any local authority attempts at design control were seen increasingly as unwarranted interference, the period 1990–92 was indeed remarkable in breaking through the log-jam of both central government and private sector hostility to the role of planning in aesthetic matters. In these three years, for example, the RTPI journal *The Planner* devoted no less than 11 issues to major articles on design control and related matters. The RIBA journal, by contrast, had only four issues carrying similar articles, but the subjects of design control and/or quality of the built environment were clearly of equal concern to both institutes. (*Planning* magazine carried several further articles on these matters over the same period).

The four articles in the *RIBA Journal* all appeared in the period May to September 1991, of which the first three issues centred around linked matters of design control and the RIBA/RTPI Design Accord. The general tenor of the architects' approach to the question of design control was immediately reflected by the May issue which bore the cover heading

'*Should planners police Design?*', and which contained several testimonies by well-known architects along the lines of how their schemes had been altered, delayed or blocked altogether by the planning process. A recurrent theme was the inconsistency of decisions and the arbitrary nature of the advice received from local authorities. A related theme also was the tricky problem of well-intentioned and helpful Planning Officers having their advice overturned or stalled when the applications in question reached the planning committee of lay members.

This rather one-sided 'discussion' in the May 1991 Journal was in fact the lead-in to the National Architecture Conference named 'Survival, Cities, Celebration' held later in the same month. A debate on Aesthetic Control, as part of this conference, took place in the Jarvis Hall of the RIBA's headquarters in Portland Place on 23 May. It was not long before the architects and the planners came into head-on collision, and the sparks flew on both platform and floor.

Peter Rees, the Director of Planning for the City of London, opened the session by saying that planning is all about pragmatic compromise, and of 'realising mediocrity out of awfulness'. (It has frequently been cited as 'the art of the possible'). He pointed out that most architects would be out of work were it not for development control, in that most developers/clients were looking to reduce costs by 'eliminating all the frills', at the same time as overreaching themselves in terms of wanting too much from the application sites. He said that planned cities, however, were cold and uninteresting to British eyes (e.g. Haussmann's Paris), and said that 'classical wallpaper'[53] over all kinds of structures was deplorable, though popular with the general public and with Prince Charles (supposedly, though the Prince now appears to have distanced himself from the 'stylistic' approach). Development Control could eliminate this kind of thing. He felt that the architects should concentrate on the main expressions in the urban scene, and the planners should 'heal the wounds' in between.

Michael Manser, the other main platform speaker, was violently opposed to any form of design control, which he felt should be abandoned. He complained that *it was too indefinite* (*'too few rules and too much opinion'*) and felt that planning should concern itself only with aspects that were measurable, e.g. land use, density, access, height and privacy. He considered that development control was essentially negative, and that planners should really stick to forward planning, which is what they were best at. Despite the fact that some consensus was emerging between the two institutes at policy direction level on guidelines more carefully worded than hitherto, which would allow architects greater design freedom, many unsympathetic/antagonistic architects' comments flowed from the floor, of which the following were typical:[54]

- 'Aesthetic advice is often inconsistent – can local authorities be sued, or incur costs on appeal, for this?'
- 'Dismayed at lack of understanding of both architectural design and planning by planning committees.'
- 'Aesthetic control exercised without the necessary skills – no longer any real dialogue – cause of delays.'
- 'Architects should have the opportunity to explain their schemes before the committee – there should be architects' panels before every committee.'
- 'All conservation and design posts in local authorities should be deleted – they do more harm than good.'
- 'Royal intervention is now a chilling underlying factor – a very pervasive stage has now been reached in this respect.'
- 'Planners should avoid aesthetics.'
- 'The general public is often more understanding than conservation/amenity groups (who operate from a biased standpoint).'

TOWN PLANNING FRAMEWORK

- 'At the end of the day, planning committees are the main, sometimes the only, problem.'
- 'Planning abhors contrast, and adheres rigidly to harmony and uniformity – this leads to mediocrity (and kills innovation or even fun).'
- 'More design education should be available to both the general public and to committees.'

Town Planners in the audience took a different view, however:

- 'Architecture is a public art, and planning committees represent the public. Architects only represent their developer clients, and themselves.'
- 'There should be more dialogue – Planning Officers have to mediate, and this should be recognised.'
- 'There are not enough architects in the RTPI – the RIBA has failed in this, and instead has either encouraged or condoned an adversarial situation between the two professions.'
- 'The planning system should be used properly, including design control, which is politically essential. The appeals system should be speeded up.'
- 'Design has to be a whole process – aesthetics should not be fiddled with, but design is a material consideration in planning. Architects should be prepared to submit a statement of design intent, or something similar.'

But:-

- 'Too many planners are trained to have a grey, negative, bureaucratic approach, and do not appreciate that the British environment thrives on the fun/muddle approach to building design and layout' (an architect).

In summary, it was agreed that the RIBA/RTPI should go further to examine the constraints within the planning system. It was generally felt that:-

- The system was vague and inchoate, but it was the product of the British disease of wanting exceptions and changes to everything.
- Continental countries and the USA operate a more structured system of land allocation, with well-produced 'books of rules' issued by the Planning Authorities to give comprehensible guidance on development, but without overly restricting the freedom of the architect to design in an interesting and innovative way.
- Planning Officers should not have the power to turn away or reject applications on design grounds alone.
- Architects should have free access to committees.

It was claimed that only two schools of planning in Britain[55] have courses in Urban Design, and that these were optional only – a very unsatisfactory basis for dealing with proposals where design was a 'material factor' (to quote the DoE's own words). Schools of architecture give urban design much greater prominence – reflecting a traditionally different approach.

The 1991 RIBA Architecture Conference was however upstaged within 2 months in the shape of the RTPI's high-profile Planning Awards ceremony at the Glaziers Hall in London, at which Prince Charles spoke as Patron of the Institute, in his first engagement in that capacity. The Prince of Wales gave strong support for planning, and for a more positive role for planning authorities in design control.

Prince Charles took the view that Planning Officers' and planning committee members' fussiness about details in design reflected a healthy concern. 'The face of our country, whether in our buildings or in the way we treat the countryside, is moulded by small, incremental, but essential decisions about detail', he said. In his view 'The essence of civilisation lies in attention to detail ... We have to devolve some power over these decisions to local level, to the level at which towns and villages are experienced.' He felt that local 'constructive protest' should not have to rely on the chance existence of local expertise in design matters. The characteristics of the local vernacular could just as well be established and recorded in a good local design guide, he said. The Prince commended in particular the North Norfolk Design Guide, and that of the 'imaginative planners' of West Dorset District Council, with whom he had had many productive discussions about Poundbury (and who also were incidentally most helpful and co-operative in the formative stages of this book).

The Welsh Brecon Guide and the one in preparation by the Scottish Countryside Commission were also singled out for praise by Prince Charles. 'Lots of sterling efforts were going on locally', he thought, 'but there also required to be central government recognition and encouragement of these efforts if they are to be given the authority they need to stem the tide of unsympathetic, arrogant development.' Government advice, through policy guidance, 'could encourage local authorities to promulgate such guides as part of, or in association with, the planning process.'

The RTPI President, Peter Fidler, said he was delighted with the Prince's public support for a design course to guide both architects and planners, and for local design guides which took account of traditional local styles.

Herein, however, lies the spark which from time to time ignites the highly combustible material contested by the Michael Manser school of thought, namely that of *including prescriptive rules on style and materials, as well as on scale or proportion, within general planning guidance* on density, massing, height, means of access, etc. By this route, we come back full circle, and abruptly, to the major underlying conflict between 'high-tech' modernism and vernacular traditionalism - *and to the unwillingness of the more militant or outspoken members of the RIBA to be restricted by the planning system (or by the planners) from doing just what they want.*

Other Initiatives: the DoE's Quality Campaign

Two further initiatives in the field of urban quality deserve special reference here – both resulting in perceptive, well-illustrated reports (one recommendatory, one consultative).

The first is the Institute of Civil Engineers' report *Tomorrow's Towns* published in 1993 from the collective findings of a joint working party of engineers, town planners, landscape architects (each representing their own institutes), plus representatives of the Civic Trust and the National Rivers Authority. The report is both general and specific in nature, with a strong practical bent overall, and contains numerous recommendations on many (essentially tangible) aspects of managing and developing the urban environment. The aspects comprising the report are set out under:

Positive Action
Coping with Traffic
Greening the Town
Making the Most of Water
Keeping up the Quality
Town Centres
Residential Areas
The Local Authority – a key player
Pulling It All Together

The second initiative is that of the Department of Environment itself: *Quality in Town and Country*. The publication of this report in July 1994 (again, copiously illustrated, somewhat in the style of *Tomorrow's Towns*, and to the typographical standard of *A Vision of Britain*) ushered in a nationwide consultation on its general propositions for the obtaining of greater environmental quality, which were set out under the following headings:-

Town and Country
Towns as Villages
Vital and Viable Town Centres
Our Local Environment
Heritage

TOWN PLANNING FRAMEWORK

Better Quality Buildings
Traffic and Travel
Housing
Quality and Investment
Taking Decisions in the Round
What Next?

The results of this consultation exercise were considered at a symposium (in the QE2 Conference Centre in London) opened on 12 December 1994 with an introductory speech by the Secretary of State for the Environment (John Gummer). The symposium was attended by more than 300 leading figures in the fields of development, planning and architecture. In this speech, the Secretary of State emphasised that responsibility for the built environment was widely shared and that concerted action was crucial, outlining his own commitment to quality across a wide field.

This was followed in March 1995 by an address to the Civic Trust by the Secretary of State, laying further emphasis on many of the points he had raised in the symposium. Particular issues addressed were:

- a 'holistic' approach;
- the vitality of towns and town centres;
- the Thames Gateway including a need for a strategy paper covering the Thames from Greenwich to Hampton Court Palace and beyond;[56]
- street enhancement, and those who damage it – noise nuisance, as between neighbours (new legislation planned to deal with this);
- the reputation of the developer;
- household projection implications;
- housebuilding provision and character;
- mixed-use planning;
- urban design;
- sustainable development.

The DoE's Quality Initiative continued in 1995 with an Urban Design Campaign, whose aims were to try to generate more debate locally about what makes for a good or bad environment; to raise awareness of the design of the local environment; and ways in which it might be improved. The Campaign was designed to test a number of the themes and ideas which emerged most strongly from the responses to the Quality Initiative. These were:

- the need to re-establish mixed use as the norm;
- the value of collaboration in the development process between landowner, developer, local authority and local people;
- the benefits of consulting the community during that process;
- the value of distinguishing clearly between urban design and architecture, in what was currently an overly style-dominated debate; and
- the pivotal importance of appropriate use of good development and design briefs.

EDUCATION FOR URBAN QUALITY

We have seen how the media and publicity play an important, if not central, role in bringing issues of urban quality into the public eye, and how in turn the outcome of controversial planning and development cases may well be influenced by the degree of public exposure and debate at the time.

Just as important, and even more fundamental, is the role of education in environmental matters, including the fields of Architecture, Planning, History and Sociology.

Design Awareness of the Lay Public

In order that informed debate and intelligent consideration take place on such issues, it is of course necessary that the participants have some clear awareness or grasp of the criteria and processes involved in resolving the issues at hand. This applies to all participants – public as well as professional, lay as well as expert. It applies with equal cogency to amenity societies, planners, elected councillors, architects, engineers, surveyors, developers and their commercial advisers, and last but not least, to the general public itself.

Clearly, in most cases, the appreciation of successful urban quality or good urban design will be gained from media dissemination plus actual experience (whether first-hand or via forms of publicity), rather than by formal education. But at the same time, there is ample scope for a wider knowledge amongst professionals of each other's field of expertise, which could come via their training. And there is certainly scope for others involved in the planning and development process to obtain a wider picture (and thus a clearer grasp) of the processes and criteria by which urban quality is obtained.

Planners, for example, need to have as wide an appreciation as possible of the architectural and urban design aspects of development schemes which come their way, and some architects certainly need to appreciate the wide range of issues plus the various difficulties faced by both Planning officers and their committees in dealing with their proposals.

Councillors and the public need to be increasingly aware of those issues that transcend the purely grass-roots aspects of planning applications, if real urban quality is to be achieved (or safeguarded). Amenity groups must similarly recognise that there is often a whole range of relevant or contributory matters which go to make up a balanced solution, in addition to the more obvious physical or visual aspects.

Financing institutions, including building societies, must come to realise that developers (via their own architects, surveyors and commercial advisers) have to give due weight to the assessment of planners and the public in gaining approval to development schemes, and to the fact that the overtly market-led regime of the 1980s now has to give way (at least in part) to the policy-led culture of the local planning system and to the growing public pressure for better development in a better environment.

Again, much of the necessary wisdom thereto is bought via the direct (and often hard) experience of the development/planning interface, via costly appeals or other delays caused basically by the failure of developers/applicants to realise the nature of the planning forum in which they are operating.

Local reactions are always relevant, and it has to be said that sheer weight of objection will nearly always have a marked effect on local authority deliberations, though not everything can be adequately resolved that way. In the most frequent context of the interface between aesthetics and planning control, the RFAC study *Planning for Beauty* showed that the parrot-like repetition of slogans such as 'It is all a matter of personal taste', or 'Beauty lies in the eye of the beholder' is not a resolution of the aesthetic issue but an avoidance of the problem: 'Worse, this undermines the traditions of high aesthetic standards and the seeking of objective standards of truth and beauty in the arts which is the foundation of our culture' (Lord St John of Fawsley, Chairman of the Royal Fine Arts Commission). The RFAC report demonstrated that such a populist response is not only philosophically and intellectually inadequate but does not accord with the facts. 'A society that lists buildings for preservation, designates conservation areas, and selects other areas as being of outstanding natural beauty, is clearly declaring its belief in objective standards', it pointed out.

If, the public, elected councillors, amenity societies and other lay bodies are to go beyond the ordinary 'subjectivist' view, then the role of the media is of particular importance in conveying up-to-date information, thinking and commentary upon development proposals of any significance or relevance, in both a modern and historical context. This will tend to be the chief, if not the only, channel through which the layman can gain a wider insight into modern development aspects, other than that afforded by what takes place locally, i.e. in his or her own backyard. One exception to this is of course where members of the public, especially within amenity groups, already have some direct background knowledge, skill, or special interest founded upon reading or university education in subjects that bear upon the quality of the environment. This, though, will more often than not tend to be of a traditional/ historical context, such as found in some members of preservation societies, whose work is of sterling value but whose horizons may need perhaps to adjust to the times in other respects.

Another exception is that of the planning committee member, who has the opportunity to attend various 'councillor information' courses, such as the annual Councillors' Summer Schools held by the Royal Town Planning Institute, in conjunction with the yearly Main Summer School held for its professional members. Other 'training' sessions may be held from time to time, or arranged, by the parent local authority.[57] Conferences, such as those held by the RTPI or other institutes, may also be relevant, depending on whether those institutes are prepared to open their doors to non-professionals/ non-members. The media, however, remain the most powerful educators of the lay public in planning, development and amenity matters generally. Publications such as Prince Charles's magazine *Perspectives* (from his Institute of Architecture) are well placed to play an important role in this way, and certainly fill an obvious gap.

Design Education

On the professional front, a particularly cogent issue is that of design content in planning courses. Planners are frequently viewed by architects as being by nature only two-dimensional, and certainly much of a planner's training traditionally has been along either one- or two-dimensional lines, although design or development briefs are quite usually expressed in three-dimensional terms, i.e. with reference to height, mass and scale of buildings.

There has been a certain amount of controversy and head-scratching within planning education circles over the question of design content in planning courses, and certainly there is no set or consistent pattern. The RTPI has its own Educational Guidelines, plus visiting panels which control the administration and running of planning courses, and it is for them to ensure that there is adequate design content. (There have unfortunately been a number of 'casualties' in recent years, in the form of planning courses closing down owing to economic constraints, with lecturers being thrown back into the local authority planning 'mainstream' and having to be found alternative scope for their career aspirations).

One undergraduate course in Planning, which is an example of successfully promoting Design as an integral part, is the Queen's University of Belfast. This course not only aims to teach students about design – it sets out to make them into designers. It was developed in full consultation with the RTPI, and was well received by the visiting panel. No members of the panel were designers by training, but they displayed no bias against a course that placed the development of design ability high on the agenda.

Other UK educational institutions which (in 1995)[58] ran separate courses in Urban (or Civic)[59] Design, in addition to their courses in Urban/Environmental Planning, were as follows:

Oxford Brookes University (School of Planning)

University of Liverpool (Department of Civic Design)

University of London (Bartlett School of Planning, UCL)

University of Manchester (School of Architecture)[60]

University of Newcastle upon Tyne (Department of Town and Country Planning).

A total of 43 educational institutions in the UK ran courses in Town Planning, with or without distinct specialisms, including the five above, as at September 1995. It is probable that a number of these included Urban Design aspects (though specific figures are not available).

Continuing Professional Development (CPD) is now an important (some would say essential) adjunct to a planner's career.[61] Unless planners keep themselves up to date, for example, they could find their professionalism being put in question at public inquiries. Local authority quality control and assessment of departmental performances also increase the need for training. The RTPI's Continuing Professional Development rules required in 1995 that members complete 50 hours of professional development over a 2-year period. Although CPD can embrace home-based learning, teaching, secondment, or supervised research, the most popular training route is the short course. And although a Planning Officer's attendance on a short course[62] not only means course fees but also absence from the office and thus more work for others, the local authority will be wise to accept this as a good investment. The same will apply to the private sector.

What of architectural education of the architects themselves? It is here, in particular, that the 'Great Divide' between the traditionalists and the modernists emerged initially. It is true to say that most architecture schools have, to different degrees, encouraged broadly the modern approach as predominant, and indeed have been in many respects both the incubator and the 'flagship' of the Modernist movement. In this, Walter Gropius and the Bauhaus School of Design in Germany during the years before World War II, together were especially influential in the development of modern European and American architecture.

The architecture schools recognised early on that a design technology and philosophy based solely within the confines of 'traditionalism' would never suffice for the needs and conditions of the 20th century, let alone the 21st century. The upward path of modern building design has never been easy, as there has always been a dedicated band of traditionalists ready and willing to oppose the development of new forms of architecture 'on the ground' at every turn. Yet new types of buildings have been built quite regularly in Britain, notwithstanding the opposition (variously) of the public and the planning authorities. Few would have thought, for instance, that the Economist building in St. James's Street, London designed by Peter and Alison Smithson in the 'New Brutalist' era of the late 1950s/early 1960s would ever have been built, and moreover would successfully hold its place over 30 years later, only a few doors away from Robert Adam's masterly Boodles' Club, with its elegant Palladian fenestration and detailing, and in a street full of buildings in consistently traditional styles. Yet it has taken place, and few would now quarrel with it. (Was it allowed via enlightenment or by default, in the first place? Some traditionalists even applaud it!)

In the public sector, the architects of the former London County Council, Greater London Council, city councils such as Norwich, and County Councils such as Hampshire, Hertfordshire and Essex have led the way in the sphere of public buildings including residential flats, schools, libraries, etc., with Essex and Hampshire also blazing the trail in design guidance for developers. Alas, the flowering of local government architecture and architects had all but died out by the start of the 1990s, choked near to death by the severe general restriction of public sector building in the Thatcher years, though here and there pockets of resistance have resolutely held out, as in Hampshire, where the County Architect, Colin

Stansfield-Smith, determinedly continued his county's promotion of public architecture.

The schools of architecture, plus the more enlightened authorities in the public sector, have thus had a considerable, if not always recognised, influence upon the development of architectural design and thought in Britain. Yet the public have remained to a greater or lesser extent unconvinced that modern architecture can successfully take place without the traditional face of our towns suffering unduly as a result. *The 'Great Divide' between old and new still remains.*

The schism is typically represented by the establishment in the early 1990s of two opposing schools of architectural and environmental design. On the one hand, the Prince of Wales, following the success of his *Vision of Britain* book, TV film, and generally anti-Modernist philosophy, followed up his intention to establish an Institute of Architecture to promote traditional craftsmanship and architectural design. At the same time the RIBA, via its then President Richard McCormac, has encouraged and pursued the creation of a British Architecture Foundation, with a rather more up-to-date remit. Each is in contrast, even if not in opposition, to the other. Behind these two concepts and projects lie the serried ranks of adherents to the opposing architectural philosophies in Britain. Polarisation of this nature on the part of the public is bound to work against the interests and aims of newly formed (and older) institutions as these.

The antagonism cannot be either healthy or desirable. It reflects a split in society which may be irremediable. If, however, each of these schools of design thought were to take the broadest view of the whole field of urban form, it is by no means impossible that a partial coalescence of understanding could occur, which would be highly worthwhile in terms of 'common ground'. *But the essential factor and 'missing link' is that of environmental education.*

Environmental Education

Finally, the question of environmental education at the most formative level – that of the classroom and beyond – needs to be addressed. In this country, where education in matters of urban quality and design tend to be given a minimal profile, the important need of the youngest section of the community to acquire this at an early stage is frequently overlooked altogether.[63] One public-spirited organisation which took up this challenge was the Lewis Cohen Urban Studies Centre at Brighton Polytechnic (now the University of Brighton), under their Director Selma Montford. Under the auspices of the South East Branch of the RTPI, they published in 1988 their own *Resources for Environmental Education* directory, listing information resources available from planning departments and other bodies in Kent, Surrey, East and West Sussex. In the opening words of the directory:

> Environmental education in its broadest sense can be defined as ways of applying the subjects learned within the classroom to the environment outside. Art and design, mathematics and biology are some of the less usual subjects which can be linked to environmental work, in addition to geography and history which have traditionally formed its core. . . . the introduction of GCSE in schools, with its greater emphasis on project work and on the development of pupils' skills in seeking for and using material, has increased the need for environmental education resources. In addition, education authorities are increasingly looking to meet the need for project work within their county and often close at hand to the schools.

The Lewis Cohen directory[64] was prepared in order to complement the work being done by teachers and local authority planning departments, libraries and other sources, in the hope that this would add to the knowledge and material already available from teachers' centres and specialist advisers, and also in the hope that it would suggest new topics to be studied.

Although regional only in its scope and

applicability, such a document is clearly of value in promoting via school teachers an awareness of the environment amongst the young, and encouraging them to have an open mind and a positive approach to it. The efforts of study centres such as these, whose 'outreach' is in fact to all sections of the public, can only be beneficial and should therefore be supported by all means available. Local efforts 'on the ground' to promote an understanding and appreciation of local environmental issues, sowing the seed of first-hand experience of urban quality in the minds of citizens of all ages, is worth a whole library of government legislation.

We have examined the joint aspects of publicity, controversy and education, following our review of the machinery of the UK planning system, relative to the provision or retention of urban quality. We shall proceed to examine under the third (and last) main part of this book, 'Perspective on Urban Quality', the context and means by which urban quality should be achieved, in terms both of individual developments, and of the public realm generally, with case studies thereto. We also discuss the fast-emerging concept of urban sustainability, plus pluralism, new technology, etc., in preparation for the 21st century.

PART II – SUMMARY/KEY ASPECTS

CONTEXT FOR DESIGN ADVICE

- In Local Plans, there should be general policies relating to desirable standards of design for new development, for conservation, for areas needing environmental improvement, and for additions and alterations to key/listed buildings in Central Areas and other sensitive areas. Local plans, while incorporating both general and special policies on desirable standards of design, should also identify sites or areas where formal planning/design briefs will be necessary.

- Local planning authorities may well need increasingly to use the services of retained development and/or design consultants where sensitive environment/major proposals are involved. But this should not beg the question of service direction internally in the full professional sense, and there may be a problem for resolution where Chief/Senior Planning Officers become unduly involved with management aspects of their Departments, at the expense of urban quality matters.

- The need for pre-application negotiations is underlined by the continued constraint on local authorities of the 2-month statutory period (dating from 1947!) for determination of planning applications (3 months where trunk roads are involved). These periods are frequently exceeded, and it is naturally on the doorstep of the planning authority where the opprobrium will lie. The saving grace is perhaps that more developers now realise that good attractive design not only helps to speed the process and to secure planning permission, but is also a good form of investment.

CONSERVATION AND PRESERVATION

- Where conservation areas are concerned, unbridled private development has resulted in a 'protectionist/legalistic' attitude by local authorities, reinforced by an 'elitist/exclusivist' view of conservation areas by the more favoured sections of local communities – an undesirable tendency, especially where in competition with disadvantaged areas for limited local authority manpower and financial resources.

- The virtues of retention and refurbishment of older buildings become increasingly evident in a climate of high building costs and energy and resource conservation. A policy of active conservation and/or enhancement has many real and positive benefits. (An obvious beneficiary of conservation is tourism.)

- A source of extra concern is the damaging effect cumulatively of many and various changes to unlisted buildings which do not require planning consent (the English Historic Towns Forum has now grasped this nettle!). A general widening of detailed control over unlisted buildings and 'minor changes' in conservation areas is requisite, beyond the powers inherent in Article 4 Directions.

- Proposals submitted for planning consent/approval should give more information e.g. via 'an environmental statement' including an examination of the urban context and the relationship between new and existing buildings, plus written and drawn evidence justifying the purpose and effects of new development, particularly in conservation areas (this should in fact benefit the applicant).

- There should be restrictions placed upon the growing number of designated conservation areas, in line with the continual restrictions placed on planning authority finances and staffing and for other reasons – a balance must be struck with other planning and environmental needs.

THE DEVELOPMENT PROCESS IN CONTEXT

- The quality of presentation to planning committees, whether in the form of written or drawn submissions by the applicants, or of oral or written reporting by the Planning Officer, is of considerable importance. Schemes should always be represented in their widest context for assessment and decision. Inadequacy of presentation, in terms of lack of clarity and/or comprehensiveness, should be legitimately treated by planning authorities as a form of 'insufficient information', as to the granting of planning consent.

- The appeals procedure is a central feature of the British planning system, but is far too heavily relied upon. It is costly and time-consuming, and tends to be cynically exploited. It can often be said to be 'an admission of failure' also, on the part of either or both sides. The British system compares unfavourably in this respect to those of other Western European countries, whose planning systems have a more clear-cut basis for dealing with applications. It is however 'democratic', and we are doubtless stuck with it.

- Refusals, and subsequent appeals, can sometimes be avoided by legal agreements to secure 'planning gain' benefits locally. These are only valid though if they are needed in the context of, and as a direct outcome of, the proposals, and if they confer benefits on the actual locality of the development scheme. Good design should not be seen as some form of 'planning gain' – it should always be a basic expectation/requirement.

FURTHER MEASURES FOR GUIDANCE AND CONTROL

Land Use/Floorspace

- The extra pressure upon central urban and residential areas caused by the inevitably tight maintenance of Green Belts means that we now have an urban environment increasingly under siege. The situation in those areas therefore calls for as great a degree as the planning system will reasonably allow of local control and regulation of all kinds, consistent with urban quality.

- The two main underlying problems are the operation of land values and market forces. Perceptive and responsible local land-use planning is accordingly needed as an antidote. Additional tools for this might include a revived use of the FSI (Floor Space Index) in simplified/restricted form, where applicable.

- A particularly intrusive element in the traditional scene is the high building. Local authorities should therefore prepare/possess a High Buildings Strategy, and/or a Building Heights Policy, where they expect unusually intensive pressures for development over a small or limited area. However, the 'floor-space quantum' must be closely controlled and restricted against such problems. The need for controls over high buildings also underlines the need for an 'urban design concept' for sensitive areas.

Other Measures for Urban Quality

- The most damaging factor in the modern-day environment is that of vehicular traffic. An essential consideration for economic vitality is that of traffic accessibility. But there is also an essential balance to be struck between the needs of traffic to flow through (or round) an area on the one hand, and the needs of accessibility for local terminating traffic and parking on the other. Extraneous traffic should be discouraged and routed elsewhere.

- Vehicular plans should be supplemented by parking plans, and traffic-calming measures wherever appropriate, and always augmented by pedestrian provisions, utilising traffic management techniques wherever possible/necessary. The need for new road construction should be minimised, existing primary road capacities maximised, and the local impact of traffic minimised, by such measures.

URBAN ENHANCEMENT MEASURES AND PROVISIONS

- Quote: 'Unlimited, unguided private activity is as destructive as unlimited socialist public control and interference'.
  (J K Galbraith, 1992)

The answers for urban/environmental enhancement will increasingly lie in joint action and funding between the private and public sectors, and in the parallel need for involvement of the local business communities as well as the general public.

- It should be axiomatic that 'good design equals good investment' in virtually any context. Poor design is money wasted and usually engenders problems for the future.

- Urban quality also relates to mixture and variety of uses, in terms of creating a friendly environment (a fact increasingly recognised at all levels, including the DoE).

- 'Preservation Pays': our architectural heritage should be seen as a dwindling commodity, underlining the case for reuse of old buildings wherever possible. There are numerous benefits to be gained from this.

- In environmental improvement, the skills of town management, exercised by the local authority, are likely to be a key factor. Various applications of this in practice include: planning gain, 'percent for art'/'public art', street schemes, pedestrianisation schemes, conservation area schemes, and relevant traffic measures.

- Where the problem/task is wider, such as in run-down inner-city areas and other areas of high unemployment, government or other grant aiding is usually necessary.

IN PURSUIT OF URBAN QUALITY: PUBLICITY, CONTROVERSY, AND EDUCATION

- No longer are matters of architecture and planning the sole prerogative of architects and planners – the combined attentions of the public and press have ensured this change in attitude. (The press/media, however, have an obligation to be accurate as well as engaging or controversial!)

- Various publications in the planning and

architectural field over the last 30 years or more have been noteworthy, and some have been greatly influential. Gordon Cullen's *Townscape* from the 1960s, and Francis Tibbalds' *Making People–Friendly Towns* from the 1990s are two examples. The one that in recent times gained the most popular appeal at its appearance was Prince Charles's *A Vision of Britain* (book and TV film), with its '10 principles for good design', immediately striking a sympathetic chord with the general public. But the images and predilections which the Prince invoked in search of good design and acceptable development are essentially rooted in the past.

- The Prince of Wales's well-publicised, and thus well-known, personal interventions in contentious development sagas as the National Gallery Extension and the Mansion House (Mappin and Webb/One, Poultry) site have revealed, as nothing else has, the unsatisfactory nature of such proposals and the circumstances surrounding them. These controversies highlight the profound split that still exists between 'old and new' in Britain to this day, in terms of architectural design in particular.

- Francis Tibbalds (who also produced his own set of 10 commandments and 14 principles for 'user-friendly' environments) got closer to the root of these problems via his four main suggestions for 'getting it right', namely:

  1 greater commitment to good design on the part of central government in this country;
  2 radical changes in the training of professionals concerned with design of the urban environment;
  3 the need for urban design to be properly recognised within local planning authority structures; and
  4 the need for a sensitive and responsible approach to the after-care and management of places.

Two apposite quotes:

> 'There is above all a need for sensitive town management to fill the role of a major catalyst of the quality of life and of our surroundings'
> (attributed to Alfred Wood, 1992)

> 'Urban design is not some kind of magic dust to be sprinkled on to dismal environments . . .'
> (1992 tribute to Francis Tibbalds)

- The British public tend to view their planning system as vague and ineffective; but if so it is probably the product of the British disease of wanting exceptions and changes to everything (we see this as synonymous with 'democracy'). Continental countries and the USA operate a more structured system, with well-produced, comprehensible and consistent 'books of rules' issued by Planning Authorities by way of generalised guidance to developers.

- The vexed question of design control in Britain is unlikely to go away. The Prince of Wales in 1991 gave strong support for planning, and in particular for a more positive role for planning authorities in design control, saying that their view of this reflected a healthy concern, and that there should be some power over these decisions at local level, at which towns and villages are in fact experienced by people. (Subsequent government guidance, via Planning Policy Guidance Notes, has however qualified this approach.)

- The role of education of future professionals in environmental matters, including the fields of Architecture, Planning, History and Sociology, is of fundamental importance to urban quality. The media, however, remain the most powerful educators of the public in planning, development and amenity matters generally (especially via TV and the local/national press).

- A particularly cogent issue is that of design content in town planning courses. Planners are frequently viewed by architects as being by training and nature only two-dimensional. The advantages of inter-professional teaching and training are evident. There should be more understanding between architects and planners: this would benefit both public and private sectors. There is also a parallel need for environmental education in schools generally – the foundations for appreciation of good urban quality and design have to be laid at the earliest possible age!

PART III

PERSPECTIVE ON URBAN QUALITY

A WAY FORWARD: INVESTMENT IN URBAN QUALITY

In what has now inescapably become a capitalist world, wholly dependent on financial returns that are able to stand comparison with many alternative deployments, investment in the environment, both natural and constructed, will always gain the support of perceptive democratic societies. For it is the quality of these environments, be they rural or urban, which provides the background and the matrix to our life experiences, and the education for civilisation of all our young.

New and often radical ways of looking at the financing of our towns and cities, and earning adequate return on such investment, are overdue in Britain today. As in so many artefacts produced for the convenience and enjoyment of society, there will always be a desire, and even a hunger, for quality at an affordable price. Our business competitors have repeatedly shown us how to apply this dictum, which our Industrial Revolution taught to the world, and it is high time that the guidance of our urban design quality should take on the energy we see at every hand being applied now to a sustainable natural world.

Investment in existing urban centres in Britain today is probably a greater risk-bearing activity than it has been for the last 200 years, and it is understandable that financial institutions now look to short-term returns in the development or redevelopment of our towns.[65] This has, of course, a profound effect on the urban quality we have come to regard as being 'traditional'.

Confidence is often engendered nevertheless by thorough investigation and full knowledge of all design parameters, properly explained and guaranteed, to reduce uncertainty and speculation amongst those affected in both the long and short terms by larger-scale urban development. Firmness of political purpose must appear to be a variable in the investment equation, when a major project might well cover the lifetimes of two or three changes of government. Money is hardly ever 'saved' by delay in a well-considered development proposal, and a curtailment of such a scheme can often prove most damaging to the economic prospects of contingent social development invariably involved in the network of commitments entailed. 'There is nothing stronger than the Truth' above all is an old adage in business, and it lies at the heart of successful design guidance towards the achievement of worthwhile ends in town planning.

The preparation of an initial brief for any development project accordingly involves the survey and selection of the most relevant and up-to-date accurate information available to the project managers. Material fed back from earlier similar examples, again as accurately collated and analysed as possible, will give valuable direction to the design team involved; but in

Britain such information is often not readily available, in the same way that it can be obtained in North American practice. This is a pity, since confidence and speed of successful execution are greatly improved when survey information is comprehensive, recent, accurate and truthful.

The developer, the local planning authority, and the architects and designers associated with any major project can only be as successful in its realisation as initial survey quality allows. There are many examples of completed schemes which, through inadequate economic and traffic survey information, have been realised in the wrong place, to the wrong size, and even at the wrong time to best serve a particular urban redevelopment opportunity.

The collection of accurate economic, traffic and social statistics, and their updating at frequent intervals should become a national task, broken down through local planning authorities. Moreover their availability during planning consent negotiations, to all parties, should be an automatic part of the search process.[66] In retailing design particularly, the disclosure of such useful information can only benefit the ultimate quality of the result, since all parties will be aware of the resources possible for commitment to a viable scheme. The proof or otherwise of viability, based on true and confident information of a comprehensive and unbiased kind, would form grounds for approval or otherwise at the formal application stage.

But the provision of successful development, however desirable, cannot alone rectify the burgeoning environmental problems around us. If we are to deal effectively with questions of urban quality, we must recognise first the outward manifestations which typically overlie the problems and which signal their existence and the need for action to deal effectively with them.

The motor car. The roads for it to use. The oil for it to burn. The acres collectively laid aside for it to drive and park upon, the green fields for it to conquer. All these have helped to increase another pollution: a pollution of too many people wanting to spread, and exploit their new-found awareness of the big world outside the urban, and even suburban, cocoon of the old cities. The 'Garden City' dream turned into outer suburb. The outer suburb leapfrogged the Green Belt ribbons to become New Towns with low-density pretensions. Now out-of-town-shopping and business parks served by an ever-increasing motorway system have made it possible for developers to make money not only from these new ways of living and working, but also from the tidy laying away of our old urban centres, in preservation and conservation.

Life in Britain has lost its urban quality in many ways. Out-of-town activity is putting undue traffic and development pressures on the countryside, and the built environment has not served any of us particularly well. We have lost our late-found understanding of the pleasures of urban life, and have forgotten what made our old rural life bearable, and often delightful, throughout the history of a very precious part of the world.

Our land area is too limited, and our population density too great, with its insatiable desire for the convenience and even for the necessity of personal transport from door to door, to permit British society today to adopt the urban planning ways of the Americas and the developing world. We must urgently look, in an essentially European manner, at ways of replanning our urban inheritance and infrastructures, to reinstitute proven areas of life in our towns and cities towards the cause of spending time in the company of friends, the mutual security of an easy contact with our fellows, and the healthier development of mankind.

New Directions

Although we have identified the need for investment in urban quality, concomitant with the present situation whereby the physical fabric of buildings, etc. attached to the land is now of greater significance in value terms than the land component itself, this helpful circumstance begs an equally important question – that of the essential relationship of the regulatory planning system with the development process. (It is largely for this reason that we have been at pains already to examine thoroughly both the planning and development systems in this country.)

Planning must be enabled to escape from its persistent image of being too systematised, regulatory and negative – and of being used by a defensive and sometimes traumatised public merely to stop things happening. It must also regain its previous professional reputation of intellectual construction, enabling whilst at the same time guiding and moulding developmental change. The future without doubt lies in the field of cultural adaptations, town management and the provision of schemes to embellish or augment the environment, creating perceptive and worthy settings for a wide range of human activities and aspirations.

This scenario – shared necessarily with a people-conscious development industry not solely preoccupied with profit or with optimisation of its own position as at present – was not facilitated up to the mid-1990s by a government seeking not only to marginalise local democracy, but thereby also weakening local town planning and other democratic machinery in the interests of market competition via measures such as privatisation, deregulation, or externalisation of local services. Commercialisation of democratic local processes by government is no answer to the fundamental needs of the future; it is instead a tempting and insidious confusion of direction.[67]

We are also now in danger of trying to retreat to some point or scenario in an impossible past, made from an amalgam of impressions from surroundings existing in former times. Decreasingly, it seems, do we look forward to a clearly conceived future. 'Futurism' is not extinct, however: it lives on in the improbable dreams of environments and artefacts not to be developed in our own time, but rather in some future resolution of our present problems, by the intervention of outside forces, which can only be imagined. The authors, in attempting this work, have preferred to suggest the art of the possible in presenting the accompanying material. That is to say, we have attempted to illustrate those factors that are currently inimical to the planning of good urban quality; those factors that are conducive to this process; and those elements of the latter which could be realised in a possible British future.

Urban Quality: The Embodiment of Meaning

The essential element of *quality* in urban environments is not something that can be easily measured, or even identified fully, as it may well spring from a combination of factors relating to 'sense of place', such as legibility, collective memory and issues of historical continuum. To these we should nowadays add 'inclusiveness' and 'diversity' in a pluralistic society. We discuss sense of place via 'identity of place' and 'urban context' principally under the later heading 'Philosophy of Context' (p. 137), so we shall not embark here upon a detailed analysis of these. What we may however point to, from the outset, is the direct emotional link we have with our built environment, via that elusive element of *quality*. It is clear that this link predominantly involves our senses, and that there is therefore *a sensory response* between our personal (and collective) psyche, and environmental form and its origins.

This leads us to identify the first and most important factor in urban quality – that of meaning. The element of meaning in our surroundings is essential to our psychological well-being. A meaningless environment is the very antithesis of what we need and expect our urban surroundings to be. Without meaning, we are lost, and life is nothing. Absence of cohesiveness, lack of clarity, purpose or structure (illegibility), or absence of evidence of origin and manner of growth (whether organic or formally designed) are all manifestations of a meaningless environment, to which human beings can only react and respond negatively and uncaringly. A meaningful quality of environment in our towns and cities is thereby the essence of urban quality. But how are we to achieve this?

First, we may draw many valuable lessons from *history*, in terms of circumstances and solutions. In the end, all problems of urban change and growth, and of their relevant physical/design components, come down to the same basic set of considerations; social, economic and cultural imperatives, plus administrative, locational, historical and technological attributives. History is littered with examples of these, either singly or in combination – usually the latter.

Second, and arising from these considerations, we may examine the role in today's world of the three main arenas for achievement and implementation of urban quality, namely:

- planning;
- development; and
- people.

Within these, third, we have the operation of planning policy/control and urban/civic design; whilst underlying these we have the twin (and often rival) factors of people power and the resolution of economic forces. These factors are, of course, reflected in the interplay of the public, private and 'voluntary' sectors – in turn a reflection of our increasingly pluralistic society.

Today our best hope of achieving urban quality lies in the beneficial use of *urban design*, linked increasingly to the concept of *mixed uses*,[68] thereby supplying the imaginative but also realistic approach needed for the creation of true urban quality.

It is too limiting to regard urban design as little more than a pragmatic, problem-solving activity; the sensory aspects inherent in environments of quality demand something more fundamental. Essential to what we need and expect as to urban quality is, as we have said, the perceptive embodiment of meaning (see, for example, Norberg-Schulz 1975: 425, 428, 434).

Our problem here is that the British town planning system historically is structured and orientated principally along administrative and legalistic lines. In today's market-led development scene, the system is increasingly required to recognise (and lend itself to) economic arguments and imperatives. Only recently has overt attention been given by government to the equally demanding imperatives of achieving and securing quality environments, for example via the 1994 DoE discussion document *Quality in Town and Country*, as well as via recent PPG Notes and Circulars (see Part II, introductory section). For all this, the British planning system does provide at least a partially effective and wide-ranging set of controls and guidance for obtaining or safeguarding aspects of urban quality, which we have attempted to highlight. It seemed to us that our study should define and describe these at an early stage, first as the statutory platform that we all need to know about; second, as a 'springboard' for considering what changes of direction are necessary for a new mode of 'administration' of the future built environment, which will be appropriate to today's swiftly moving and changing world.

It is certainly true to say that 'quality' of environment will not change unless the more restrictive or outdated aspects of the planning process are improved. New policy approaches to land use are rapidly displacing existing planning procedures. These policies are at the embryonic stage, but the need for a fully integrative environmental policy[69] which can be applied across the board and which embraces many disciplines is already recognised at all levels of government. It is generally now perceived that planning will have to operate in a qualitatively different manner in future, i.e. via new organisational and policy procedures, incorporating to a greater extent the views of ordinary people (never to be underrated), plus greater co-operation between the public and private sectors.

PHILOSOPHY OF CONTEXT

Virtually at any time in history, some form of positive planning has been axiomatic to the successful achievement of urban quality, whether by local prescriptions as to types, amounts and mixtures of uses permissible in a particular locality; by encouragement of specific building forms, normally related to a local range of materials; by control of floorspace quantum and scale or heights of buildings; or finally via the process of conscious creation of visually artistic and civilised layouts in which buildings are assembled in the particular type of relationship with each other and with the streets and spaces complementing and responding to them, which collectively fall within the scope of urban design.

In all such situations in which urban quality is achieved, one common and indispensable factor must in some way be present: the recognition of the identity of the local urban context (what we would more simply call 'sense of place') as the essential basis for designing and providing forms of development. This can be achieved in one of two ways:

- by adding to it 'organically' by way of infilling or augmentative structures, to change that scene 'in kind' and by degree, or
- by entirely replacing outdated and inadequate structures or layouts in creative, imaginative, yet compatible ways.

Unless there is the accepted remit to wipe out or totally ignore the existing features/characteristics of a given area (an increasingly rare circumstance nowadays, except in areas of industrial dereliction or residential squalor), the observance of local urban context must be the essential first step towards accommodating urban development or change; all else in terms of urban quality will flow from this.

The response to 'urban context' will most usually be in terms of harmony[70] with existing building forms or layout, though not necessarily implying slavish detailed adherence to historical styles. Alternatively, circumstances may in fact warrant contrast rather than harmony (the resolutely modern Pompidou Centre, in the Beaubourg area of Paris, adjoining the largely 17th- and 18th-century Marais quarter, is a prime example of radical contrast). There will be cases such as this where the existing 'context' is inadequate to genuine and overriding needs, and thus where a new scale or 'key' must be set by new development for a successful future urban scene.

Sometimes the opportunity for innovation will be given by the very richness and diversity of the inherited urban fabric – sometimes justification for total change will arise from the absence of any worthwhile or redeeming features in the urban scene. But it is always the question of identity of place, in whatever degree it exists in terms of range of uses and their physical expression, which should 'set the scene'. Hopefully it should guide and inform the developer, urban designer, or town planner towards a design solution that is compatible with, rather than simply imposed upon, a valuable heritage, or which overcomes the deficiencies of the existing context.

There are innumerable instances throughout history of the successful application of this basic principle. With some notable exceptions, urban places are not normally created in one self-contained 'tranche' or 'episode' – rather, they evolve. Historical evolution, above all, is the process by which identity of place is derived, however cohesive or otherwise that urban character is. That evolution includes the present

Old Bristol

PERSPECTIVE ON URBAN QUALITY 139

Figure 16 Bristol: historic and maritime legacy.
A city with a great maritime trading past, Bristol must now look to the future, whilst conserving and making best use of its superb natural setting and its splendid architectural legacy. The 19th-century lithograph (opposite) bears witness to its historic importance as a commercial and cultural centre.
  Despite its busy commercial life though, an air of uncertainty seems to lie over its quaysides and historic precincts. New uses are needed increasingly for open waterfronts and neglected old buildings if Bristol is to remain true to its distinguished past, and if urban quality is not to be forgotten.
Sources: Photo, Gordon Power; lithograph of Bristol (detail) by J. Lavars (1887): City of Bristol Museum and Art Gallery

day, of course, and the assessment must also include all the various physical results (past and present) of expediency, haste, cost-cutting, uncertainty of purpose, and over-ambitiousness, as well as the manifestations of foresight, fitness of location and purpose, sensitivity, and a feeling of consistency, balance, reasonableness and proportion throughout – rare commodities in these days of maximisation of development opportunity at the expense of cohesiveness of urban character. So the urban context which we have inherited via the process of historical evolution is almost invariably an imperfect chequerboard, providing both constraint and challenge to those who would deal with it.
  History, as we have said, provides us with countless examples of our 'philosophy of context'. It is worth considering some of these, especially where they vary significantly. At basis, several factors are in common – the social needs and uses that prompted and influenced the settlement and its subsequent evolution; the physical nature of the site and terrain itself; and new or variable economic circumstances. Of these, only that of the physical siting of the settlement has a constant bearing, though this too will be subject to both opportunities and constraints as the original settlement area develops outwards, and creates new land requirements, and thus new patterns of living and activity.
  Excellent examples of 'development in context' from past times are the Georgian terraces of Bristol, only a few miles away from those at

Bath, which similarly exploit their hilly site. At Bristol, the majesty of the Clifton Gorge, with its 19th-century suspension bridge by Isambard Kingdom Brunel, heightens the drama of the urban scene, while the city simultaneously embraces its historic docks, in close proximity along the Avon riverside area. Other examples from the urban heritage in Britain of successful interaction between site, social needs, economic stimuli or imperatives, and resulting development, come readily to mind – e.g. seaside stucco Brighton and Hove, with the grandly laid out Regency squares, streets and terraces of Kemp Town and Brunswick (by the local developer-architect team of Wilds and Busby), or Edinburgh New Town, whose Classical layout of equally gracious 18th-century stone-built residences, on the plain to the north of Princes Street, both complement and contrast with the older city areas of the Castle and the Royal Mile. The mill towns of the North of England, in their robust and forthright character, are equally good examples of response to both context and need.

These historical developments, built cohesively over a relatively very short period for the particular needs of an emerging population class, are valid examples of a response on town or city scale to those needs and to the opportunities of the more typical British (and even European) urban tradition. Our urban character and context has though been shaped more by the pattern of cumulative needs of settlements than 'single tranche' development, however attractive and stirring the latter idea can be. The English country town, with its range of historic building fabric dating usually from medieval times or earlier (as in the case of towns like Chester or Chichester, each on its Roman camp gridiron plan; Salisbury, on its 'chequers' layout; or Lincoln, Canterbury and York, with houses and shops still grouped closely around the great cathedral church, as in the Middle Ages), is much more the 'role model' in most people's minds for the quintessentially English urban settlement. It is here that informality and illogicality, in the form of small-scale intricacy and 'fun/muddle' (as Francis Tibbalds so well put it), still hold pride of place over grand axial layouts, large scale, spacious squares, and other manifestations of Classical formality which we associate with the Continental home of urban experience. At the lower end of the historic urban range, we have the individual qualities of the smaller market or county towns (such as Lewes in East Sussex), displaying a richness and diversity of character and often an array of local styles and building materials which many an overseas visitor considers to be unrivalled and enviable. Environments such as these must continue to receive the greatest care and sympathetic treatment, though still enjoying (within their own scale and 'key') a freedom to create visual forms of no less interest or validity than those which over the centuries have combined to create that same variety and liveliness which we so often value.

Clearly, the amount of change that can take place without 'swamping' the urban context, or causing irrevocable loss of character, will depend not only upon the amount and type of development proposed, but also upon the scale and nature of the urban context for which it is intended. Larger towns and cities are nearly always more 'robust', and thus more suitable to receive developmental change, than their smaller counterparts. The latter increasingly tend to suffer from the paradox of needing on the one hand to maintain an adequate degree of economic activity to avoid stagnation (or worse), whilst on the other hand requiring great care in the handling of incremental change, and certainly needing protection against radical change (though the latter is not normally the problem especially in periods of recession).

The most insidious and often the most damaging influence upon the urban context, is that of traffic and its demands in terms of movement, access and 'storage' needs. With the increasingly heavy imbalance in favour of road transport and use of private cars

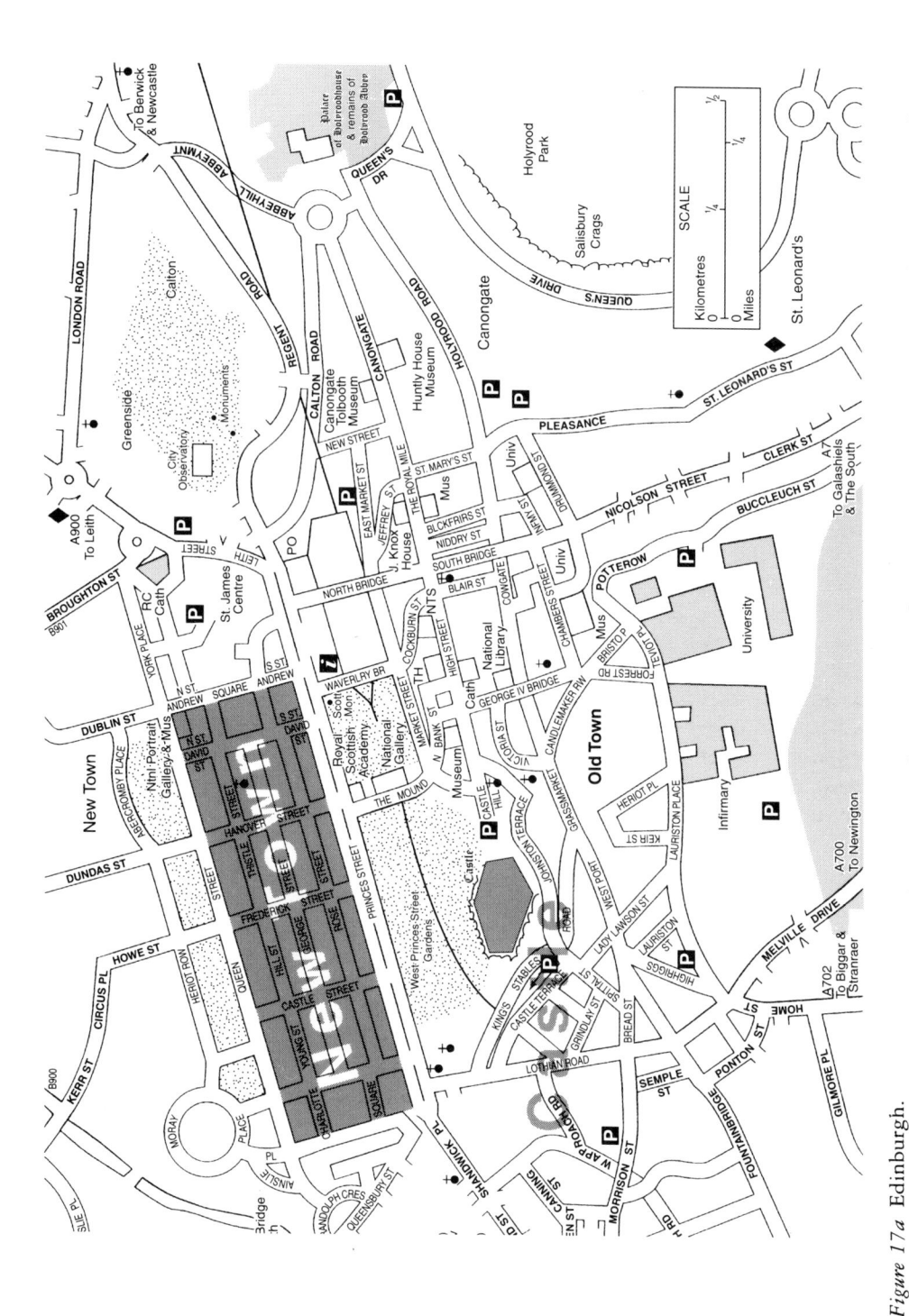

Figure 17a Edinburgh.
New Town Conservation Area: 'A planned urban concept of European significance, the New Town has an overriding character of Georgian formality. The First New Town, built to James Craig's 1767 plan, has experienced significant redevelopment, while the Second, Third and Fourth New Towns, which were laid out on estates to the north, east and west retain most of their original buildings. Stone-built terrace houses and tenements, built to the highest standards, overlook communal private gardens; to the rear are lanes with mews buildings, many of which are now in housing use.' (Written Statement, Central Edinburgh Local Plan, September 1994.)
Map source: O.S. street plan

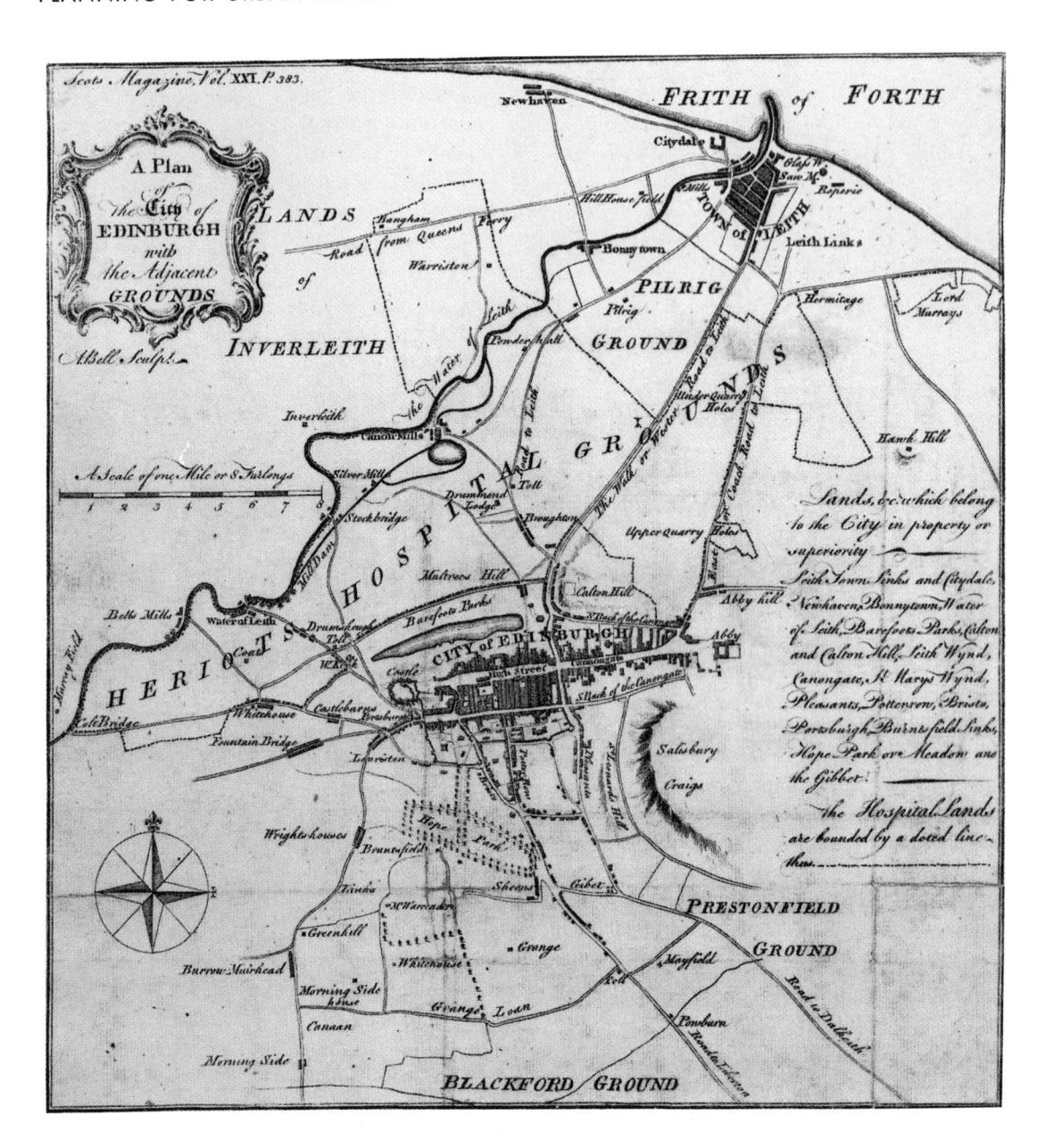

Plan of Edinburgh and Adjacent Grounds engraved by A Bell for *The Scots Magazine*, July 1759. This small print was derived from the large-scale survey made by Fergus and Robinson in 1759 (*RCAHMS*)

Figure 17b Edinburgh: New Town and Castle/Old Town.
Edinburgh's Castle, New Town, Princes St Gardens and Calton Hill are in classic combination through the natural setting. Enhancement of the Castle could be gained by relocating the annual Military Tattoo to a new site in Princes St Gardens.

PERSPECTIVE ON URBAN QUALITY 143

In the New Town, modern office requirements are now incompatible with some 18th-century buildings, and residential is regarded as the preferred use in these cases.

Priorities
The Vision for Old Town is articulated in a series of proposals and projects, ranging from an enhanced visitor experience at the Castle to a new plaza forecourt at the Palace.

Sources: New Town – *James Craig, 1744–1795*, Mercat Press, Kitty Cruft/Andrew Fraser and *Planning Week*, 27.7.95, RTPI. Castle proposals: LDR, Columbia, Maryland, USA (as against public transport), and the creation of a growing amount of major road network (in line with what is now happening all over Europe), the direct implications for accommodating this surge in traffic and roadspace (especially in south and south-east England) relative to the maintenance of urban and rural quality is highly disturbing (with forecasts of doubling of road traffic by 2001). Environmentally, with the effects of pollution added in, traffic growth is without doubt the single most adverse factor which has to be taken into account in the regulation and safeguarding of our urban contexts. There are strong arguments for the augmentation of Highways Authorities everywhere with environmental/urban design units to cope adequately and to deal sensitively with the increasing pressures of traffic upon both the urban and rural environments, not merely to carry out 'surgery' to accommodate more and more change, but also to resist either the incursion or the generation of extra traffic pressure (especially the latter) where it will result inescapably in damage to the urban context. Alien or incompatible forms of building proposals are sometimes more easily perceived and dealt with than the insidious and often ruinous effects of unrestricted traffic growth and its accommodation within our urban environments.

In different places, the ability to introduce new development safely will vary considerably with regard not only to visual aspects such as scale, style or other architectural aspects, but also in terms of use and activity generation, especially of course traffic and parking. *There are a great many urban areas in Britain, however, which do not fall easily into any single category, being essentially of mixed use, scale, layout, period and character* (many in fact the legacy of the 19th century and latterly of the equally uncertain conditions arising from a changing and receding late 20th-century economy). It is most often these urban areas which require both economic change and physical change, in a variety of ways and forms.

Modern building forms are seen as less sympathetic on the whole to the established/inherited urban context, though normally more directly suitable to accommodating modern commercial or similar uses, than our traditional architectural forms and building methods. It is felt by many that the more modern technological building forms are most suited to 'stand-alone' schemes or island sites, and that their integration into traditional townscape, whilst not to be entirely ruled out, can present degrees of difficulty not previously experienced. Schemes using traditional forms, materials or designs still score more heavily with both the public and with planning committees. An example of this recurring theme occurred via a joint RIBA/district council design competition held in March 1994 for a prominent and historic riverside warehousing site in Lewes, East Sussex. From the many imaginative entries, the adjudicators, who were two distinguished architects (one local, one of national standing) selected 'modern technological' schemes for the first three places. Lay opinion, however, as reported by the *Sussex Express*, was affronted by the adjudicators' decision;[71] there was a strong local feeling that a more traditional type of design should have won, although the locally-favoured scheme (which directly echoed the nearby 18th- and 19th-century warehouses) could equally have been said to be a fairly straight case of derivative historical styling, if not actually 'pastiche' – a familiar story!

The 'principle of context' thus applies to any assessment of development, virtually irrespective of size of proposal or scale of urban context. The determining aspect in the end will be that of quality – i.e. quality of proposal relative to quality of urban context – thus emphasising and endorsing the whole question of urban quality (the theme of our book) as the central and pivotal matter. An alternative way of expressing relationship with urban context is that of 'contextual compatibility'. *Environmental Aesthetics: Theory, Research, and Applications*, (Nasar 1992) has examined the question of 'contextual compatibility' in building, to determine whether or not this is basically a simple matter of personal taste. Two groups of lay people and one group of members of design review boards were asked to assess 25 buildings in terms of site planning (defined as mainly the positioning of the building relative to its neighbours), massing (i.e. height, bulk, scale, etc.), and style (mainly in terms of facade detail, though this concept might be regarded as a trifle limited); and to rank these on a scale between the two aspects of 'contrast' and 'replication' (i.e. harmony). Interestingly, a high degree of accord and consistency was found between the two (lay) groups. The most significant result to emerge was that the key factor in people's assessment of compatibility was facade detail, rather than site planning or massing considerations. Most importantly, compatibility of detail between neighbouring buildings did not in their view imply simple replication but could embrace a fairly wide degree of variation and innovation in architecture. It may be that similarities of facade detail offer a more tangible basis of assessment of compatibility than other aspects, where lay people (and thus the general public) are concerned, or it may be that this is regarded as evidence of compatibility in a 'family appearance' sense, whilst otherwise allowing variation of shape, size and layout. The danger is that of gaining general coherence at the cost of undue restrictiveness of design, where planning control is concerned.

We shall look at various aspects of urban quality, in addition to that of 'compatibility with context', and suggest various examples that embody such aspects. First, we should consider the question of what makes for 'development quality' (in terms of successful and viable development) and then go on to look at the question of the public realm in planning, design and development, and how this much-maligned and neglected aspect of our urban environment is increasingly valid and relevant today.

DEVELOPMENT QUALITY: COMPONENT FACTORS

There are manifold considerations (apart from key locational needs) in any given situation regarding the extent to which urban quality is the outcome of architecture, layout, spaces, juxtapositions of all three of these, mixtures of social activities and land/building uses, plus prevailing or past economic conditions. We shall need to look at certain case studies and other examples, to assess urban quality in terms of these various factors, and also to see how both old and new developments interact with urban context, either successfully or otherwise.

One particular aspect (relative to location) which needs to be understood, but is often not fully appreciated by the critics, is that of *the set of principles by which development itself is conceived and proposed* – and also whether these criteria necessarily lead to what we would wish to call 'development quality'. By the latter, we must therefore mean the principal ingredients for successful new development,[72] at least in the developer's eyes, but hopefully also in a wider context. These can be set out as follows:-

- **Finance**, including provision for maintenance;
- **Amenity**, i.e. attractiveness for people and uses;
- **Traffic and parking**, including provision for pedestrians and access;
- **Security/safety** for people, uses, property, etc.; and
- **Order/organisation** i.e. clarity of purpose, whether applying to single or mixed uses.

They probably require little explanation, except for the comments that follow.

Finance and Maintenance

All development is essentially a risk venture, where capital and economic activity are invested in a building project against the prospect/aspiration of worthwhile/substantive financial return, plus enhancement of value of both site and the buildings, etc. upon it, over a forecast period. Aspiration may run too high in some cases, leading to conflict with Planning Authorities (and public), and often to overdevelopment and resultant problems such as excessive physical scale, traffic generation, detrimental effect on other land uses/economic patterns, etc. if the development proposals are approved (e.g. on appeal). But without the willingness of a developer to 'compete in the marketplace' and to run a risk, little or no development would come about, whether for good or ill. It is again worth noting that in today's development economy (in Britain at least), there has been a noticeable swing, whereby the value of land is now taking a secondary position to the increasing value and significance of the buildings attached to the land. One factor, however, still tends to be neglected in the UK – that of suitable financial provision at the outset for adequate future maintenance of the buildings and their sites. This omission is probably more often the result of 'accountancy economics' on the part of finance houses, as much as any cheeseparing tendencies on the part of developers or their advisers. (It should also be added that, compared with well-designed developments, poorly designed ones do not last as long, require more maintenance, and are valued less.)

Amenity

This may be defined in any number of ways, but 'attractiveness for people and uses' as regards either old or new development is surely axiomatic for good development: it is central to the whole matter of urban quality. If a building or development in the above terms is attractive to people, it will normally mean not only that people enjoy its presence (i.e. they like its appearance and feel it is right for its location), but also that they enjoy using it (i.e. it meets their needs and is convenient both in location and accessibility, and also in terms of design, layout, concept, etc.).

Traffic, Parking and Access

On the whole, developers will pay more attention to the needs of access to, and parking upon, their site than to external traffic movement, including the creation of additional highway loading caused by their proposals. However, the Highways Authority will normally not be slow to remind the developer/applicant of the latter problem, and in most appeal cases it would certainly form a major aspect of submitted evidence. In some cases, a developer may even offer to contribute financially to the improvement of nearby roads, via the 'planning gain' machinery, to help secure a planning permission. This is perhaps a measure of how far things are changing, as the traditional view of developers has been that provision or improvement of roads is something they would certainly wish to benefit from, but have no great wish to contribute to.

Adequate access (both vehicular and pedestrian) to a development site should be of approximately equal interest in a developer's eyes to the efficiency/sufficiency of the general traffic movement in the area. In practice, questions of access are far more relevant to the developer than provision for traffic external to the site. Inadequacy of access 'in degree' benefits nobody, of course, whilst provision of vehicular access for service, visiting and parking needs, so far as these apply, may be so fundamental that it will normally form the subject of a 'ransom strip' situation where it requires the crossing and thus the acquisition of a piece of someone else's land (or an easement over it).

Pedestrian access to sites and buildings must equally be fully provided for by an intending developer, from a public footway system. Moreover adequate arrangements now need to be made for disabled persons' access to all new buildings visited or used by the general public (a provision not to be neglected).

Security and Safety

In these times of mounting vandalism and crime, security and safety for people and uses in developments of all kinds have climbed high up on the agenda. The prudent and responsible developer must now give far more attention to, and put a lot more resources into, these particular provisions. Crimes against property and persons are still principally found within the major urban areas (but they often also occur nowadays in quiet country towns, where they might be least expected). This will sometimes affect development in a quite visible way,[73] e.g. in security measures to be taken in shopping malls in some major town centres against 'ram-raiding', in the form of bollards or barriers, or against other forms of intrusion, such as by the use of closed-circuit TV monitoring and of steel shutters. Similarly in housing estates, where regard now must be had to questions of providing adequate sightlines and of eliminating 'blind corners' where miscreants can lurk, and of introducing sharp thorny bushes and sill-height growth as part of an estate's landscaping, to deter intruders. The solution very often lies in careful detailed layout design of these developments, and local police forces will frequently give advice at the planning/design stage, via applications upon which they are consulted, as to security measures which can be taken generally.

Equally, all kinds of detailed safety measures must now be taken by responsible developers for the protection of the public, and these are controlled (and enforced) by local Environmental Health Departments, often quite stringently.

Order and Organisation

Lastly, order and organisation in the design of a development are clearly essential to its success. This applies not only to how well it functions in itself, but also to how well it relates to its site. This will be a matter both of careful planning and skilled architectural design – the two are integral. How the development relates in turn to its urban context is a matter of equal importance for the achievement of urban quality.

The need to provide for single or mixed uses on a site may well result in different types of architectural solution. The most modern building technologies, e.g. those employing space-frame construction, or large areas of glass or metallic/plastic-coated cladding, are sometimes better suited to single-use projects, rather than to mixed-use needs.

Single-use projects are probably easier to design than mixed-use schemes, in terms of internal organisation and order. However, where they employ modern architectural methods and visual forms (whether for functional, economic or prestige reasons), they can be more difficult to relate to a 'mixed character' or traditional urban context, unless there are good reasons simply to 'impose' the development on its context. Those reasons might include either the self-contained nature of the site (e.g. at Stanstead Airport, or at the 'new' universities, such as East Anglia, Lancaster, York, Keele or Sussex); or the tight integration of the site within a macro-scale context (e.g. the Lloyd's development in the City of London); or the degree of relevance of the urban context to the scheme in terms of use or scale (e.g. at railway termini such as Waterloo, Charing Cross, Liverpool Street/Broadgate). A good example of a large modern office building 'imposed' successfully, as a freestanding block, is the pagoda-like American Express building in Brighton, very much a landmark feature, in an urban context devalued by banal tower blocks (see Fig. 11, p. 67).

One might add further examples of buildings with an individual and/or prominent location in extensive grounds, and therefore with a 'proprietorial' nature (e.g. Chester, Epsom and Sandown racecourses, Lord's cricket ground, or Glyndebourne opera house). At many of these types of location, all essentially 'single-use' in character, it has been possible to realise individual buildings both in tune with their surroundings and to a high standard, extended in truly modern form. Whether such buildings dominate or merely 'fit in' is probably as much a matter of circumstance as of intention.

Mixed-use schemes are generally more difficult to design than single-use projects, especially where the range of uses has to be closely integrated into one overall development layout, together with their differing service access needs which usually have to be combined. (The Ashley Shopping Centre in Epsom, Surrey, is a good example of this, where a theatre plus town housing, and new offices for a petroleum company, all had to be 'fused' into a major covered retail scheme in the town centre.) Integration of new development on a large scale into an existing/historic urban context is never easy, but with mixed-use schemes opportunities are sometimes available to fit them satisfactorily into the existing scene by expressing their different elements and juxtaposing them relative to different parts of the urban context which at the same time have a corresponding scale or character. Additionally, there may be opportunities to build a large amount of new development at least partly behind a major/historic street pattern, providing new servicing, parking and access simultaneously above or below ground level as necessary, on a common planned basis. This, of course, both assists in preserving the street scene, and in fitting in a large piece of development unobtrusively but with full functional requirements integral. The Ashley Centre in Epsom, and the more recent 1990s Castle Mall shopping centre development in Norwich,[74] are both excellent examples (in their different forms) of the above principles.

THE PUBLIC REALM: NEGLECT, DECAY AND REVITALISATION

When we speak of 'the public realm', we mean for the most part those elements of the built environment in Britain and abroad that are common to the general public's collective use and enjoyment – a wide and perhaps loose definition, but at the same time covering an area much too long neglected, and now deserving of rather closer attention and a greater degree of action and support than hitherto. Additionally, of course, we should consider to an extent the ability (or otherwise) of the public sector agencies, in particular the local planning and Highways Authorities, to contribute to the provision of the 'public environment'.

Both the public realm and the public sector have, at different stages, simultaneously undergone fluctuations of perceived importance and therefore of funding support, at least in Britain. Here, the period following the end of World War II (i.e. the 1950s and 1960s) was marked by a need to cater for the housing, social and employment needs of a rising population, at the same time replacing worn-out or obsolete buildings, and providing for traffic and pedestrians in ways that would largely meet their requirements for the foreseeable future. For this 'brave new world', the modernistic buildings of Scandinavia in particular, and of modern architects like Le Corbusier, Alvar Aalto, and Walter Gropius, formed role models in many respects, including the use of geometrically simple (and often severe) buildings faced in concrete, glass and steel, disposed about or rising out of pedestrianised precincts, around which traffic was planned to flow, with parking and service areas woven in. Examples of this 'new-look environment' were Coventry Central Area and the Rotterdam Lijnbaan (both of which replaced bomb-devastated urban core areas), and the British New Towns, such as Stevenage, Harlow, Crawley, East Kilbride and Cumbernauld, which were largely built on open/rural land, but close to or contiguous with existing small country settlements, which then became districts of the wider urban area. New central shopping areas and 'neighbourhood centres' thus arose, and pedestrian areas were accordingly created. Straight from the drawing-board to the site, it all looked very clean, imaginative and exciting at the time.

Thirty or more years on, the intervening period has seen the outmoding and deterioration of many of these public realm developments, owing partly to the need in the 1990s for more sophisticated forms of living and environmental facilities; partly to the largely unforeseen social problems inherent in high-rise council housing schemes (some now unfit or dangerous to live in); partly to progressively changing modern requirements in commercial and residential developments; and partly to lack of adequate maintenance provision. The whole brave new experiment in a modern living environment for the 'new age' has perceptibly gone sadly adrift, brought about by a disillusionment and loss of confidence, not only in the forms and concepts of the day, many of which have undergone recasting or even replacement (e.g. the Bull Ring at Birmingham or Churchill Square at Brighton), but also in terms of severe curtailment of the previously recognised role of local authorities in providing a public realm environment.

By the beginning of the 1990s, there had been a huge 'sea-change' in this type of provision, away from the public sector and towards the private sector, which took place progressively in Britain over the 1970s and 1980s. But, despite the swing towards private sector provision, it is still largely unresolved as to how the public realm will be catered for.

A major problem, largely unsuspected back in the 1950s or even the 1960s, has been the unremitting growth of vehicular traffic and its demands, pressures and effects upon our urban areas. One might perhaps argue that this situation should have been foreseen by the government (or those responsible for it) at the time. In fact it was: the 1963 Buchanan report *Traffic in Towns*, to the Minister of Transport, was 'a study of the long-term problems of traffic in urban areas', and was highly perceptive as such. In this, Britain poses a particular problem – although the substantial level of vehicle ownership here may not be the highest in Europe, the level of use of cars in Britain is one of the very highest. A typical local traffic survey, taken at a peak hour anywhere in the UK, would show that about 80% of vehicles were 'driver only', for example.

We love the individual convenience and freedom which the private motor car offers us, but seem oblivious or uncaring as to the inevitable effects of our over-use of this vehicle in a small island such as ours, both upon our ailing public transport system and the environment of the public realm. In other words, we have ourselves largely to blame in respect of the damage to our public environment inflicted by the level of use of the motor vehicle. Another major share of the blame should of course be allotted to successive governments who have consistently (and more than tacitly) supported the 'roads lobby' and the car manufacturers at the expense of public transport, in which there has been a progressive failure to invest or subsidise the necessary operational infrastructure.

By 1994, however, the reluctant but nevertheless misguided policy of central government to keep adding to major highway systems such as the M25 Greater London orbital road (thus drawing in more and more traffic upon it and into the metropolis), and also indulging in various other overambitious road building schemes in various parts of the country, had begun to wilt and crack under the pressure of mounting public objections and financial costs. What clearly was needed was a much wider spreading of the money available for roads into more modest but adequate local network improvements, linked to local economic needs (for example, the Kent/Sussex strategic road 'box', related not only to the M25 and London, but also to the South Coast, and making better provision for connections to important but historically under-resourced areas such as Newhaven).[75] There is a need to bring Britain's road network more in line with the rest of Western Europe, but not to the extent of 'overkill' (which has a highly detrimental effect upon both our urban areas and our countryside), and not in terms of continued downgrading or neglect of much of our rail network, when in fact an integrated transport system is the primary need. In Britain, however, these lessons are slowly and painfully learned.

Traffic is not the only problem area deserving our attention, where a combination of lack of foresight or imagination in the past (or at present) on the part of both central and local government, with a shortsighted approach also by the private sector, has resulted in damaged, inadequate or plain uninspiring local environments. *Some of the circumstances which the public realm in Britain has suffered from and in some respects continues to do so, can be set out in summary as follows, with their urban quality implications*:

UK Circumstances/Context	'Urban Quality' and Other Implications
(A) *The need in the 1950s and 1960s for a 'brave new world' of widespread post-war reconstruction and social provision.* Use of Continental urban design concepts (e.g. those of Le Corbusier and Scandinavia), linked with compulsory purchase powers and special financial provisions and arrangements for local authorities, etc. Resulting in: ● Comprehensive development areas; high-rise council housing; single-use zoning; redevelopment instead of conservation, including creation of new 'instant' public realm environments via above means.	● Over-simplified/visually stark/unduly uniform buildings, in facing materials with poor weathering potential for a damp, non-Mediterranean climate as in Britain, and thus maintenance problems. ● Marked or stultifying uniformity of visual environment disliked in comparison eventually with much more 'local-scale', humanised forms/designs.
(B) *Transition to 1980s saw failure generally of 'comprehensive' approach* in both development and planning quality/results, as evidenced opposite. (NB seen as failure of public sector planning and architecture)	● Social problems associated with high-rise living in tower blocks (isolation, vandalism, crime, physical deterioration of surroundings), or similarly with empty, windswept, hard or soft pedestrian areas, not greatly used by/unpopular with the local population. ● Devaluation in real estate terms with buildings either vandalised, closed up, or often now redeveloped as inadequate or undesirable for needs and standards of today (e.g. offices, flats). ● Loss of older building stock, where lower-density schemes could preserve such.

UK Circumstances/Context	'Urban Quality' and Other Implications
(C) *The 1980s and early 1990s saw failure of market forces alone* to supply urban quality where and in forms most needed. The legacy of the 1950s and 1960s era (as above), and the transition of the 1970s, passed into the commencement of the 1980s bullish new culture of thrusting private sector action, coupled with the progressive emasculation of the public sector (powers and finance), i.e. replacement of local democratic systems by overt preference for the operation of market forces. After 1979 General Election, under Conservative administration. local authorities/town planning both now struggling to come to terms with growing market forces situation. *Merging into imbalanced hybrid position*, where urban quality dependent either on enabling role of local authorities, or on prescience of private development sector, or on joint action by public and private sectors to provide adequately for the public realm, in the 1990s and beyond.	• *Redevelopment instead of regeneration of town centres* where needed to create new commercial areas; plus encouragement of poor or mediocre buildings (either modernistic or pastiche) and general lack of attention to townscape elements of public environment, unless part of a major redevelopment scheme – merging into: • *Selective approach to upgrading of existing building stock and local environment*, according more to market forces dictates than to any other factor; problems of reconciling profitability with broader planning aims. Need for better basis of provision of urban quality.
(D) *The 1990s continued to illustrate the failure of market forces alone* to provide a humane and successful environment as is apparent from the widening scenario of urban neglect and social deprivation. Realisation of lack of policy direction in planning for urban quality relative to general patterns of development investment provision in Britain. Upgrading of public realm, in practice mostly dependent on new development and need to relate this adequately with social context. In larger towns and cities, an issue of the increasing privatisation of the public realm, and the implications for social control that go with it, particularly in respect of shopping centres.[76] Perceived failure of both the public and private sectors *separately* to treat the public realm in an adequate and satisfactory manner, including the vital basic provision of town management and development funding, logically suggests joint action for the 1990s and beyond.	*Enlightened new local authority planning concepts, attitudes and practices jointly with the private sector* resulting in increased recourse to:- • Phased implementation areas. • Low-rise, variable-density housing. • Mixed town centre uses. • Conservation of historic areas. • Enhancement of historic areas. • Prestige central commercial uses. • Prestige B1 commercial/industrial 'edge of town centre' estates. • Encouragement of high-quality design, both modern and vernacular.

UK Circumstances/Context	'Urban Quality' and Other Implications
New agencies for development and/or its funding progressively created, as need for property investment becomes more evident, post-recession. But creation of 'quangos' to replace local authority roles and participation in development and other fields is both undemocratic and unhealthy.	• Increased attention to context, townscape, spaces, materials and durability, detail and scale. • Increased use of development or planning briefs as guidance. • Increased sense of structure and cohesion in urban areas/town centres, etc.
Under severe pressure, local authorities rise to the challenge, within financial restraints imposed, by developing improved professional planning service with new concepts, attitudes and practices, including increased use of consultants, and in-house advice/guidance generally to developers and applicants.	

THE PUBLIC REALM: A HISTORIC BASIS FOR MODERN URBAN DESIGN/QUALITY

Introduction and Review

A meaningful study of urban design, however selective or brief, that is 'complete' (i.e. holistic and relevant to today's world and tomorrow's scenario), must widen the scope of the term to include *underlying economic and social factors* that are integral to the organic/dynamic way in which urban areas change and grow. The urban designer must embrace and understand these factors, or the end result will be cosmetic/superficial, with a limited validity. The most successful town planning and urban design schemes of the past have been founded upon this truth.

As an introduction therefore to the subject of urban design in modern terms, we commence with a broad review of the socio-economic investment imperatives essential to achieving urban quality nationally. To summarise the position, central government in Britain has failed to ensure the necessary supply of fiscal energy on an evenly spread basis, to produce the required level of improvement in our urban environments.[77] The authors propose that a *regional basis of development funding*, linked with strategic planning, would be a more satisfactory and effective way of achieving urban quality, via localised urban design generally. In short, we need to reinstate the type of *'provincial' approach to quality of life,* via the built environment, that has so often met with eminently successful results in earlier times.

Before we examine the role of urban design in creating, improving or maintaining urban quality generally, and also look at the related questions of architectural design and technology in our urban areas, an expansion of the current background considerations already referred to above is worthwhile, if we are to retain a broad view and comprehension. It might be stated in summary that much of the question of urban quality boils down to a repudiation of the coarsening results of powerhouse economics in favour of a more even distribution of funding over the whole country, wherever necessary bypassing the City and central government, and locally based big businesses. Investment in the British urban fabric might also be attractive to foreign investors, keen to lease and refurbish a neglected gem of an 'Offshore Island of Europe'!

In the DoE discussion document of August 1994 entitled *Quality in Town and Country* much was done to outline and clarify the urgent need, presently existing in Britain today, for both the enhancement and sustainability of a much higher quality in the built environment. It states that urban and rural planning and architectural design were in the forefront of elements in our surroundings needing a much greater investment of professional skill and financial input than perhaps has ever been contemplated by government since the 1960s. In its very laudable aims for massive improvement in our cities and villages over the immediate future, examples of worthwhile achievement were put forward as touchstones for an early realisation of high-quality environment, across a Britain securing economic growth and expecting greater wealth through an improved quality of life in more pleasant surroundings.

Over 20 of the 30 or so illustrated examples of urban and rural designs had been funded though out of national or large-scale public company resources. One particularly controversial scheme being realised at Poundbury in Dorset was shown twice in the discussion document, as being not only a good example of traffic and pedestrian integration in its layout, but also a model for community consultation essential to achievement of the new intentions. Poundbury had in fact had a chequered reception from its promulgation by the Prince of Wales, and during its design phases by the European town planner, Leon Krier.

No firm analysis of how to provide the massive funds needed to improve or even maintain our built environment in the UK, was put forward for discussion, other than the expectations that three government agencies would have for high quality despite the limited financing available. Here the document was seriously deficient, and as in many previous attempts by central governments since 1947, the unwillingness or inability of national administrations to set the supply of fiscal energy needed to produce the rapid improvement in urban quality across the majority of our towns and cities and through our changing rural life is signally apparent. The very delicate balance between cost and the value gained for sustainable development of a high quality requires the re-education of a public taste and will so as to be capable of deploying distributed wealth in a manner more akin perhaps to a Walpole than to either *dirigiste* central governments, or those putting forward the ideas of market forces, whilst maintaining an even stronger hold on the free play of the significant amounts of investment capital, so essential to building.

The *realpolitik* of both urban and rural realisation of higher-quality environments and building has always, in modern terms, rested with the general marketability of small- to medium-scale investment, over at least three generations. Government spending is driven by the political need to continue in power. Large-scale private development is almost wholly controlled by a stock market geared to institutional shareholders, who must see a return on invested capital in under 10 years, possibly even sooner, for a speculative involvement in environment and building. It has been no accident of history that democracies since antiquity have proved to be inflationary, and a radically new approach towards the appreciation of the real fiscal value of quality throughout the life of so-called 'permanent' construction in the urban environment must be urgently re-examined. The expectations of increasing populations in the developing world must be modified, as in the so-called 'developed' societies, by the deepest understanding of resources decreasing in real availability to virtually everyone. The lifestyles of 19th- or even 18th-century northern hemisphere populations cannot be the model, subject to a little stylistic change, for immediate future generations to emulate. It will be problematic to preserve even partially our existing achievement in high-quality environment created by our ancestors across the old world.

The way forward even to attempt the regeneration of our planning for urban quality must very soon lie in the most urgent re-examination of the way we wish to live, with very limited resources of all kinds, in the immediately foreseeable future.

A lifestyle can be enjoyed, and even improved, by a common involvement in a robust and rigorous, well-engineered, 'no frills' environment. We have very little time left to reorder our tastes and aims, in a present atmosphere of conspicuous and often totally irrelevant and damaging consumption of disappearing resources, to design and install the truly new quality, which will be worthy of long-term and therefore lower-cost financial investment. Many educated young people are fully aware of the challenges, and are keen to find solutions that work. If the lead cannot come from national and international policies soon enough, any attempt to promote erratic change from societies coming into crisis will fail in a way that could prove catastrophic to the manner in which even the most unambitious of us would care to exist.

How this new lifestyle (acceptable to most, yet viable in material terms for the world's increasing population) might display itself in an urban environment, in its many traditional functions (becoming fast redundant) is so amorphous as to defy definition. New global communication techniques, the demise of carbon-producing transportation, biomedical diagnostic discoveries, all ready for early application, literally stun an empirical assessment by educated amateurs: the raw material of elected governments and their executives. If definitions are made, action is often belated, ill advised, or even outdated and incompetent as the situation quickly changes, and legislation with action becomes confused, too little (or too much!) and certainly too late. Lack of cogent direction leads to uncertainty, to speculation, to inflation, and to eventual recycling chaos in civic affairs. The orderly refurbishment of our urban environment becomes impossible under such circumstances, and it becomes necessary to re-examine the methodologies traditional in our own way of life.

To the authors it is apparent that resources to effect change with improvement would best be deployed, and publicly actioned and accounted for, at a much more regional level than that presently contemplated by centralist governments. Waste could be much more carefully observed and controlled. Known individuals would become accountable for quality success and failure, in the realisation of environments closely chosen by the populations using the facilities so designed and implemented. National homogeneity would be visible in the universal higher quality of differing solutions, suitable to regional purposes, whether in Edinburgh, Cardiff, Plymouth, Bristol, Norwich or London. The stultifying sight of the dreary 'standard superstore' or the 'motorway business park' would be modified or perhaps scrapped by competent regional Planning Authorities. Architects, engineers, entrepreneurs, and even London financial institutions drawing on world-wide capital resources, wishing to invest in a Western European success (resulting from highly motivated regional populations) and proud of tangible environmental achievements, would all play their part. They might also find the financial returns there at least as satisfactory as those processed in the central exchanges, far away from provincial energies and innovations, in both thought and culture.

The great British cities of the 18th and 19th centuries and their closely supporting regions, produced some magnificent quality in their urban planning by very local methods of finance, in often comparable revolutionary change, with half our present British population resource. It might well be just possible to 'return to the future', to establish a new quality in urban planning, beloved by those directly producing it.

Urban Design in Modern Times

The DoE's discussion document was not the only new publication at that time to highlight quality in the built environment. A more wide-ranging approach encompassing social as well as environmental criteria, was provided by *Quality of Life in Cities*, which was an overview/guide to relevant literature from the British Library in association with the London Research Centre. Although the attempt by the Secretary of State for the Environment (John Gummer) to initiate serious debate about how to achieve good urban design represents a welcome step forward in Britain, the real point is that people are guided by a variety of reasons in choosing where to live or work – not just the outward appearance, however important that may be.

There is a danger (as with the cult of conservation areas) that the elitist taste and aspirations of the financially and socially secure may deflect primary concern for housing, safety and public services generally. It is encouraging, however, to read in the DoE document that 'good design is a civic matter and not just one of personal pride or profit', and also that 'buildings, towns and cities are the one inescapable art form' and that responsibility for their appearance needs to be widely shared. We again make the point however that, as Anthony Fyson, the Editor of the RTPI Journal wrote (21 July 1994):

> The continuing aggrandisement of economic activity leads apparently unavoidably to larger enterprises wielding the power to transform our surroundings on a huge scale. The result is a coarsening of the built environment with a loss of intricacy and humanity, and a self-defeating desire to accommodate the motor car.

He also commented that over the last 15 years (i.e. from 1979), reluctance at the highest levels of government to accept public responsibility for the public realm has bred a generation all too ready to dismiss design in the community interest as any kind of priority.

The Roots of Modern Urban Design

Towns and cities are essentially symbols of order; by building permanent urban settlements, people acted to impose their will not only on their surroundings but also on their fellow beings. The shapes of both houses and settlements reflected the order and view of man in universal terms. For example, circular settlements tend to predominate in societies that live close to nature, as in the village forms of nomadic farmers in Africa and some American Indians. Many traditions preferred the stark geometry of the 'gridiron' or 'chequerboard' for planning and organising cities. The ancient Chinese identified the perfect square as the ideal city layout. In Western Europe, the gridiron plan was usually the basis for ancient Greek and Roman settlements (such as Alexander the Great's town of Priene, Asia Minor), or in Roman military camps all over Britain (such as at Chichester), or as 'chequerboard' layouts as in planned medieval cities (e.g. Salisbury). These were consciously planned towns where order was imposed socially and physically.

On the other hand, many medieval or later towns or villages (such as the South German 'village towns' of Dinkelsbuhl and Rothenburg-am-Tauber, or the Italian, French and Spanish hill towns, or again the English or Dutch market towns such as Kings Lynn, Gouda and Delft) followed much more 'organic' lines, with narrow streets winding outwards from market squares and spaces, often centred on the church, or fringing canals where water was the basic element in the town's configuration. These 'organic' towns should not, however, be regarded as of unintentional layout – their contained, compact and functional planning was every bit as intentional as the later, more formalised layouts of the Renaissance, baroque, or 18th- and 19th-century 'planned' cities.

The most spectacular form of the 'organic' city in the world is, with little doubt, that of

Venice. Here, the combined factors of building on largely artificial islands in a marshy lagoon, alongside a network of man-made canals branching out from the great watery highway of the Grand Canal, and then developing from the original defensive settlement into the seat of the greatest seagoing Mediterranean empire of all time, has resulted in the incomparable and unique city that we know today. Even the great outdoor focus of the Piazza San Marco, with its campanile, its 'piazzetta' and its Byzantine cathedral church, is a prime example of the formal space planned, in fact, on irregular lines.

Apart, however, from the many and various examples of the 'organic' town or city, from the humble market towns of Europe to the supreme example of Venice, the more formalised layout was generally utilised as a means of imposing religious order and social control, and in order to sustain harmony, in what were to become for the most part the major cities of the world (such as Paris, Rome, Mexico City and New York, but with London as a major exception, where the present sprawling layout still largely reflects the coagulation of the original scatter of 'urban villages'). To this scenario may be added the recurring cycle of prosperity and growth, decline and decay, with cities and towns becoming more and more dependent on the outside influx of goods, services, activities and employment.

All these factors have their direct implications today for the process of urban design as an essential part of urban development, conservation and renewal. After the 18th century, Europe and North America experienced unprecedented 'urban boom' as a result of the Industrial Revolution, world trade and the growth of travel (of both short and long distance). Ever since that time, an ongoing battle has had to be fought in an attempt to counter unplanned and disorderly urban sprawl, by introducing where possible coherent planning concepts, designed to turn settlements into basically 'orderly' places.

Both formal and informal planning, intelligently carried out, can successfully create ambience (that elusive but vital element for which imagination and sensitivity to character must inform and infuse mere 'building activity'). An essential element, however, is the presence of 'theme', whether formal or informal, tangible or intangible, in order for any urban design exercise to be successful – i.e. as opposed to mere tinkering with existing urban form, without contributing 'in kind', and without regard to prevailing character.

Urban Design as 'Planned Evolution'

All towns and cities evolve: some in a largely organic way, others in a more obviously planned way. Whatever the balance between the organic and planned approaches, urban design may be regarded as a process of *planned evolution* – that is, the use of physical planning and design skills combined with a study of socio-economic factors to achieve the necessary change in urban forms in an evolutionary manner via:

- sympathetic continuation of the existing street/building format or
- radical departure from that pattern where necessitated, more usually as an adjunct to retained historic forms, occasionally replacing them entirely or
- 'total' planned new settlements (whose roots go back to earliest times), the need for which arises periodically where town extensions are not the answer.

Examples of the *'sympathetic continuation'* approach are towns such as Lewes, East Sussex, where successive additions or overlays have very largely retained and respected the historic format based on a long linear high street extending across the River Ouse gap in the South Downs, lined and surrounded by tight-knit, medieval-scale buildings, streets and alleyways or 'twittens'. Departures from this scale, e.g. industrial areas, are confined to low-lying ground, previously floodplain, or at the foot of the Cliffe chalk quarries – the only exception to this careful restraint being the 1960s curtain-wall tower blocks of the county council offices, located without compromise or sensitivity at the upper end of the town, overruling their surroundings and dominating the Downland landscape.) In these gem-like towns, urban design has to be more akin to surgery than anything else, and the price for their safeguarding must be eternal vigilance, practised by local authorities and amenity societies alike.

Examples of *the radical departure* approach are mostly appropriate to the larger cities, where economic and functional necessity dictates events and forms of a more sweeping nature, e.g. in the northern mill towns of Yorkshire and Lancashire, which grew dramatically during the Industrial Revolution from the original small towns.

As *adjuncts to retained historic forms*, we have the well-known examples of James Craig's and Robert Adam's Edinburgh New Town, the Brighton Regency layouts of Busby and Wilds in Kemp Town and in Hove at Brunswick Square, and similarly the Georgian terraces of Clifton, Bristol – though these developments were achieved under relatively favourable conditions compared to the dire predicament of 'development at all costs' in many parts of Britain.[78]

As examples of *city-scale replacement planning*, we have the baroque Rome of Pope Sixtus V, Bernini and Michelangelo, and pre-eminently that of 19th-century Paris, whose rambling and decayed medieval quarters were largely pulled down and replaced by Baron Georges Eugène Haussmann (for the Emperor Napoleon III) with 100 km of magnificent radiating, gas-lit and generously treed boulevards of grand neo-classical terraces that could be controlled by the fire-power of the army (against the then real possibility of insurrection arising out of ferment). Haussmann also modernised the city's services, including building in a modern pumped water supply and a sewerage system, followed by market halls, an opera house, main railway stations, and English-style parks. Not only did Paris become a city ready and able to welcome visitors from all over the world, but it also served as a role model for other European cities such as Vienna (where the Ringstrasse was built in the mid-19th century along the former defensive walls, by then leapfrogged by the city's growth and consequently demolished by Napoleon, enabling the later creation of the grand circular boulevard of great museums and administrative centres, set in spacious grounds), or Berlin (which grew in the 19th century from 200,000 people to over 10 times that size, and was transformed into an imperial capital by a generous layout of wide avenues, largely to the design of Karl Friedrich Schinkel, the great Prussian architect-planner of Greek Revival fame).

Of equal note, however, is the extensive 19th-century civic development of central Barcelona, by the Spanish urbanist Ildefons Cerda (1815–76). An enormous area of the city (only fully appreciated through the medium of aerial photographs, in conjunction with Cerda's original plan, showing remarkable adherence) was replanned to a gridiron layout, upon which were imposed four avenues radiating from a great central square. Cerda's layout is especially noteworthy in terms of his detailed treatment of street intersections in the grid. Instead of the more usual right-angled corners of junction blocks, Cerda bevels these off at 45 degrees, creating multiple urban focal points of character and spaciousness, catering both for traffic and for 'pavement life' in true 'civic' style. The pattern is still clearly recognisable today despite the regrettable 'intensification' infill of the original courtyard areas of Cerda's grid blocks (due to rising city land values and the inevitable pressure from developers to build upon any available form of undeveloped space). Fittingly, therefore, do the Spanish refer to Cerda as 'El hombre que reinvento Barcelona' – a description needing no translation or justification!

But the likes of Cerda, Haussmann and Schinkel, or their great works, can have no direct parallel or even influence upon European cities today, except perhaps in soullessly reconstructed, war-devastated areas of Berlin. The grand urban layouts (which can be traced back to the refined civic 'space-body' of the Place Stanislas at Nancy, France, plus the great parade ground of say Ceske Budejovice, Czech Republic) no longer have any real place in urban design apart from latter-day 'civic layouts' such as at Montpellier, France, or except where landscaping and paving can be used to renew and enliven existing neglected urban spaces. Today, we must often do our best to reutilise and adapt existing central street environments, e.g. by creating new 'footstreets' where possible, and improving or providing street lighting, tree planting or street furniture, instead of relying directly upon the building of co-ordinated architectural schemes on a scale or extent sufficient to create civic character and cohesion with ambient human activity.

New Towns and Cities

The 'grand' or 'total' approach to urban planning was to make one more outburst in Europe before subsiding almost completely. In contrast to the 'domestic' scale proposed by Ebenezer Howard's *Garden Cities of Tomorrow* (1898) for the essentially English New Town settlement in the English countryside, and the two 'Garden Cities'[79] of Welwyn and Letchworth which were then actually built on Howard's principles, schools of planning and architecture arose in France and Germany after World War I which proposed very different urban forms. Le Corbusier in France and Walter Gropius plus Mies van der Rohe in Germany saw new solutions in designing concentrated 'machine for living' blocks on a city scale, utilising modern architectural technology and catering freely for car traffic. Le Corbusier was the most avid exponent of these new forms, to the point of inhumanity; his Unité d'Habitation at Marseilles (where hundreds of people could be housed in 'cells' contained in one all-purpose slab block), and later his new capital city of Chandigarh in the Indian Punjab (echoed by the new city of Brasilia in South America) were the modernistic hallmarks of the post world war era of reconstruction and replanning. In some parts of the urbanised world (e.g. in Eastern Europe) his ideas still hold sway to an extent, but in Britain they have been rejected almost totally, following the severe social and constructional drawbacks experienced from building high-rise slab or tower blocks in which to house people.

More 'human', and successful on the whole, have been the post-war British New Towns, planned and carried out in order to relieve overcrowding in the major UK cities. Usually these were expansions of existing country towns, such as Crawley and Harlow in England, or extensions (suburbs) of cities like Glasgow (in the case of East Kilbride). Occasionally, they were new settlements on open sites, such as at Cumbernauld in Central Scotland, or Hook in Hampshire (never built). In Scandinavia the New Town of Vallingby in Sweden, for instance, is part of greater Stockholm, having been planned as a satellite town but finishing as a suburb. In America, the architect Frank Lloyd Wright's vision (Broadacre City) was based much more on the idea of 'self-determination', in terms of each household having about an acre of land, so that the individual family could grow its own food, in contrast to Howard's 'co-operative' ideas.

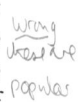

Similarly, in England, the latest manifestation of the New Town/Garden City concept is that of Milton Keynes in Buckinghamshire, where a projected population of some 250,000 people is planned to be accommodated in 8,800 hectares of land, divided up into blocks of individual ownership, related to a segregated traffic/pedestrian/cycle movement system. The low-density layout, supplemented by copious tree-planting, plus the traffic segregation, makes for a calm, residential atmosphere, but lack of a cultural centre to some extent has held back the development of Milton Keynes into a fully provided-for city in the established sense.

In spite of their shortcomings, however, and their less-than-grand domestic character, the British New Towns have shown in their way that urban design can be successfully applied even to unpretentious human-scale settlement forms, now being developed in other parts of the world, exploiting the extra freedom and possibilities afforded by informality and the 'organic approach'.

URBAN QUALITY: SITUATION ANALYSIS

Review

In this book we have progressed from an examination of the planning and development systems in the UK, including the underlying market factors, in so far as they influence or produce 'urban quality'. We have given detailed attention to the planning system, since it nowadays impinges so very widely upon most aspects of the urban environment, either as a sieve, or forum, or a battleground for the shaping of urban form and life. From this broad basis we eventually turned to the area of considering historical and ethical aspects of 'urban context' – that vital factor with which identity of place is essentially bound up. We also examined what has been happening in the public realm, that equally vital aspect of our urban environment which has suffered too much from neglect, decay and the misguided approaches of central government administrations obsessively preoccupied with constant reduction of public expenditure.

A view of historical perspective might (on the face of it) lead us cynically to believe that the urban quality we now possess owes more to our legacy from the past than to any of our present endeavours. A closer examination, however, reveals that each age has its own means of reasserting itself over time, and of utilising the best of the past in combination with the more constructive efforts and sympathetic results of the present.

URBAN SITUATIONS: RELEVANCE AND CAPACITY FOR CHANGE

In an examination of differing types of urban situation, we must first consider what types of change seem most relevant or appropriate. Change has always been inevitable in any urban situation, and it has not only to be economically beneficial (albeit a first necessity) but also at the same time conducive to urban quality.

The principal types of urban situation to be considered, in terms of settlement scales and their realistic and acceptable potential for change are as follows:

1 cities with major centres/subcentres in European and UK terms (major 'downtowns', in American terms) – but contrasted with 'inner cities';
2 large towns/smaller cities } including those of special character, plus ports, resorts, etc.;
3 medium/small towns
4 historic towns/conservation areas;
5 suburbia and 'arcadia';
6 new towns (past and present); and
7 urban fringe and *'urbs in rure'*.

In broad terms, most types of urban change will tend to occur within the town centre/'downtown' locations, or within the subcentres, for categories (1) to (3) above, with some exceptions, e.g. the old 'inner-city' subzones. In these, change may come about in a whole variety of forms and ways including:-

- **urban renewal** on a planned/'integrated' basis, with specific development plans as

well as generalised local planning policies, including certain redevelopment/demolition implications;
- **new/infill development**, either on an integrated or an ad hoc basis, occurring anywhere within areas of ongoing viable commercial or other activity, but much less so in rundown or neglected areas;
- **urban conservation**, often related to urban renewal (where funding for conservation forms part of a renewal package) or else related to permissible changes of use (where economic benefit flows from more remunerative uses);
- **redevelopment/demolition** (where renewal/infill/conservation/changes of use are not viable), arising from outworn building fabric or economically non-beneficial context.

In looking at 'relevance for change', we must also consider capacity for change. Very few places, if any, are entirely immune to change, but some places are likely to be more vulnerable to change than others. This can mean not only more subject to change but also more likely to suffer if the effects of such change are ill considered beforehand. Looking at the categories of urban situation set out above, we may draw a number of conclusions.

In (1) city centres and (2) large town contexts, there is a major likelihood or inevitability of change (due to the high level of economic activity and corresponding land values) but at the same time a raised level of 'duty of care' to secure a revitalised but also humanised urban context. In a congested and neglected inner-city situation, lack of confidence historically breeds lack of investment, loss of population and reduced economic activity, creating a downward spiral that only determined action, injection of funding, pragmatic planning, and sometimes the force of particular circumstances can correct. In (2) large towns or small cities, there is usually limited scope for major new urban developments, without risk of congestion or 'overkill'; whilst the existence of good public transport and parking is essential.

None the less, some of the older towns and cities in the English Midlands or in Central Scotland that have seen a steady decline in both economic activity and working population would welcome a return of both these components for prosperity, quite apart from questions of urban quality, where both above elements have been lost either to the supposedly more prosperous South,[80] or to the economic recessions of the 1980s and 1990s generally.

In (3), medium or small towns, the scope for large new developments is considerably less, although this will not necessarily deter the ambitious and competitive developer. Whilst he will have to accept the onus of providing adequate parking for his proposed development (to the standards laid down by the Highways Authority), he will regard the provision of an adequate road system as chiefly the responsibility of the latter.

So crucial have become the two aspects of highway capacity and land/development values that the UK has begun in recent years to experience the same phenomenon as appeared as early as the 1950s and 1960s in the USA – that of out-of-town or edge-of-town shopping centres and other commercial ventures. In Britain, these have taken the form of major urban fringe developments, of which Thurrock Lakeside in Essex is probably the largest in terms of floorspace (including mass car parking and bulk discount shopping). Watered-down versions of the same type of mass shopping development are now appearing in many suburban and edge-of-town centre sites in the UK, on redundant commercial or industrial land. Whilst this may perhaps be welcomed in terms of established or historic town centres being spared a clearly unacceptable impact upon their character, ambience and agreeableness of life, very close regard must be had to the incipient draining effect upon their economic well-being. In Britain, central government woke up very late

(1995/96) to the threat to town centres from out-of-town shopping and other developments. Planning policy guidance was issued at this time, but not before many major peripheral sites were given planning consent for hypermarkets.

In the even more car-orientated United States, the effect upon many cities has been to drain and impoverish the downtown areas as well as the inner-city areas. The danger was recognised in the USA as far back as the late 1950s, in visionary city planning schemes such as that for Fort Worth, Texas (by Victor Gruen Associates), where a peripheral road system providing partially sunken loop roads giving access to major areas of parking close to a totally pedestrianised commercial/business 'core' was proposed, in order to deal with the city's chronic congestion and economic stagnation. The Fort Worth scheme has not, however, proved capable of full implementation, despite the passage of time since its design. One of the (still) most successful English city plans has been that of Norwich, where the 'ring and loop' system instigated in the late 1960s by Alfred Wood (reflecting in UK terms the Fort Worth plan) continues to prove its success in keeping the city centre economically prosperous and viable.[81]

The Norwich city plan has proved so successful that it has been possible to build a large new covered shopping centre (the Castle Mall) alongside the Norman castle, at the heart of the city centre. To have achieved this during a period of national economic recession is remarkable, and far outweighs some of the less than generous criticisms heard from the 'Sunday supplement' architectural critics. In America, the deleterious effect of large 'out-of-town' shopping development upon commercial well-being and character has in recent years been more directly recognised and steps have been taken to reverse the trend. The Urban Land Institute of the USA sponsored in 1988 the publication of a seminal work *Designing the Successful Downtown*, by Cyril B. Paumier of Land Design Research International, Inc., Columbia, Maryland,[82] which fully addresses this problem.

In the case of both (3) medium/small towns and (4) historic towns/conservation areas, the local environment and character are increasingly vulnerable to change in varying amounts, either in the form of alien development (where the local economy is active) or decay (where the local economy is stagnant or declining). At most times and in most places, some new development is usually necessary, but again this must be 'in context', in terms of scale, location and overall character. It will be important to ensure the retention of traditional local uses in the form of a vibrant mix – the essence of real character in so many of our towns and cities, of whatever size or configuration.

Generally, the smaller the town the more 'personal' the character. Big cities can be very impersonal, and hitherto the British genius has been on the whole to personalise and humanise urban environments of all kinds, except where market forces have run riot.

We have looked at the main types or scales of urban situation likely to be affected by change, and also some types of change that are most applicable or relevant to this. Change in nearly all such situations manifests itself normally in terms of change of economic or functional role(s). This, in turn, will determine to what extent an urban area will grow, or remain static, or decline. This will then impinge upon urban quality, via the types of land use involved, the level of activity, and the physical basis of change, e.g. planned/unplanned or cohesive/incremental formats, as above. Role changes to towns or communities may be complex or subtle, as in a gradual transition of a historic town centre (such as Epsom, Surrey) to a major shopping/office centre; or they may be simple, sudden and even brutal (such as in the unplanned closure of a shipbuilding company on the Clyde, the planned closure of a coalmine in Yorkshire or Durham, or the threatened loss

162 PLANNING FOR URBAN QUALITY

Figure 18 Metropolitan/Macro scale: London/Fort Worth, Texas.
In Britain, an intelligent approach to providing a large amount of mixed-use development with a cultural content was London's Barbican scheme in the heart of the City, albeit still at macro scale (above). In America, the revitalisation plan for Fort Worth, with its traffic/pedestrian provisions (left), was echoed later in the 1960s by the 'ring and loop' plan for the city centre of Norwich (UK).

Planning and building in the 1950s and beyond in major cities of both Britain and Europe, as well as in the Americas, was characterised by high-rise development almost everywhere, either directly carried out or promoted by city authorities, often following Le Corbusier's impersonal 'Ville Radieuse' planning ethic, ignoring social needs. The Loughborough Road housing estate in London (opposite) has had a strong influence of this kind on public sector design, both then and since.
Source: *The Elusive City*, Jonathan Barnett (1986), Herbert Press

PERSPECTIVE ON URBAN QUALITY   163

[Loughborough Road Housing estate, London infl. by Corbusier]

of a county council as a traditional seat of local government in a vulnerable historic Sussex town).

Environmental Capacity

All the factors outlined above may be covered by the term 'relevance for change'. *We have to look again at the question of 'capacity for change', however, as affecting urban quality.* In this, probably the most cogent concept we can utilise is that of 'environmental capacity'. This concept is not new, and has formed part of the process of drawing up town maps and development plans, via the UK town planning system, for over 20 years.

It has recently gained individual prominence via the 1993 BDP/Arup economic study Chester – The Future of an Historic City, and we introduce it here as an important and highly relevant tool for the control (and hopefully also the promotion and enhancement) of urban quality – not only for historic towns or cities, but for urban areas generally. Via a range of representative examples, one may extend the methodology to identify issues pertinent to typical urban situations, according to role characteristics or changes, and levels of activity. Generally, the approach is that from identified situations and issues, one may identify a number of indications of 'environmental capacity' – which will then suggest the level

164 PLANNING FOR URBAN QUALITY

Figure 19 Local/humanist scale: Greenville, S. Carolina.

Following the demise of most forms of public sector development in Britain, due to her political/economic policy, increased attention must be given to any opportunities for joint public and private sector development on a local/humanist scale. An excellent example of this is LDR's scheme for the enhancement and future growth of Greenville, South Carolina, USA.

This far-sighted scheme provides for a very attractive mixed-use development at Riverplace, of which a major imaginative feature is a semi-circular central lawn both relating to and continuing the amphitheatre of a Performing Arts Centre across a river.

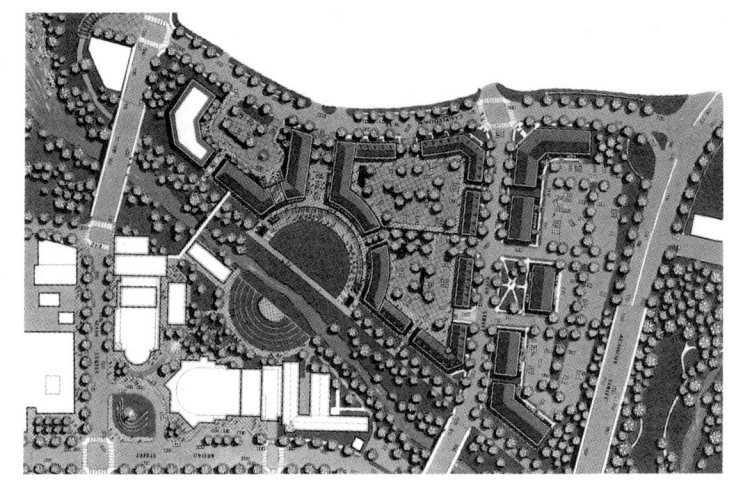

Source: Land Design Research International of Columbia, Maryland, USA

of activity or types of change that can be either acceptably accommodated, or alternatively that may be inimical to urban quality generally.

In the Chester study, a number of options or scenarios (relating to roles and activities) were looked at:

1 'little change' (implementing existing planning consents, followed by policy restraint);
2 enhancing the city's tourism role;
3 allowing massive sub-regional growth;
4 improving the quality of development investment generally (in an enhanced role as an international centre for business and tourism, and specifically becoming a university city);
5 attempting a strategy to reduce the amount of activity overall, including its retail, housing and employment roles.

These options were tested on a 'capacity framework' and an assessment of development opportunities. It was found that Chester has considerable scope for *change*, but that its *growth* could not go unchecked – and that there was a point at which the city could risk losing some of its irreplaceable characteristics, not least its historic legacy (there are 752 listed buildings in Chester, and 13% of the city is designated as conservation area), or its 'green' elements and natural setting (the city has 562 hectares, or 1,400 acres, of open space, and is surrounded by Green Belt). As a basis for the study, the *'environmental capital'* possessed by the city was defined. This included significantly:

- well-defined key edges to the urban area;
- the compact nature of the urban form of the city;
- environmental features of special importance (e.g. the River Dee and its corridor, the Meadows, Abbey Green and the Cathedral precinct);
- the townscape, plus the interaction of historic fabric, urban and green spaces, gateways and streets, all on a varied, intimate, and human scale;
- archaeological remains; and
- the quality and range of shopping (e.g. in The Rows).

The main pressures identified in Chester by the study related in particular to: the setting and scale of the city; the need to conserve the scale and fabric of its historic areas; the level of pedestrian crowding in the main shopping streets; and the growth of traffic. The main recommendations of the study accordingly related to these pressures, plus the 'environmental capital' set out above, all with direct implications for urban quality.

The Chester study notably has important relevance for other towns, historic or otherwise, in which we wish to preserve, control and enhance urban quality. The relevance lies not so much in the specific form or issues as in the methodology we may apply to those places, or to selective examples. 'Environmental capacity' is really a means of collating information about the environment of a place systematically, identifying the 'environmental priority' elements or aspects, and testing the impacts and effects of different actions. *Some of the main analytical themes which could be adopted for studies of other towns or urban areas would include*:

1 the size and form of the place/town/city, and how this contributes to character and setting;
2 the importance of public open space (hard or soft) to character of place;
3 the effect of large or tall buildings on the local townscape, historic buildings, 'urban grain', skyline and principal views or settings;
4 the quality of listed buildings and the activities which impinge upon them; e.g. modern needs and/or type/degree of use;
5 the transport network and its relationship with the character of the city, in terms of activity generated, with emphasis upon public transport advantages/benefits; and
6 assessment of development potential of sites,

166 PLANNING FOR URBAN QUALITY

CHESTER'S ENVIRONMENTAL CAPACITY

- Areas of key environmental or landscape value
- Areas with potential capacity
- Well defined edges of the urban area

Canal

River Dee

PERSPECTIVE ON URBAN QUALITY 167

Figure 20 Chester: historic city context/capacity.
In 1992 Chester became the subject of a study to determine whether further growth would harm its character or not via 'environmental capacity' techniques applied to different growth scenarios. Basic to this was the identification of areas worthy of protection/retention, covering not only the historic central core, with The Rows (above left), but also the River Dee corridor including the Racecourse (above right). From the 1994 BDP report we show (opposite) the consultants' assessment of restraint needed in the core, plus areas with potential capacity for growth.
Sources: Plan: BDP (Building Design Partnership) with Arup Economic and Planning; photos: M. Parfect

Figure 21 Bristol: RC Cathedral/Clifton Club.
Architecture of high quality, whether small or large, new or old, must be respected in its urban setting. Such buildings act as 'barometers' of the time and the mores of the societies that realise them, and they should be allowed to control their own identity and surrounding space when newer development threatens to overtake this right. Bristol's Clifton Club (right) from the 18th century and the nearby Clifton Cathedral (left) from the late 20th have this control of setting well stated.
Source: Gordon Power

within guidelines derived from a capacity framework.

It is beyond the scope of this book to enter into overall environmental studies of complete towns or cities like Chester, but we may certainly apply to selective examples a *'methodological' approach to urban quality, via 'environmental capacity' techniques* such as above.

ISSUES AND INDICATORS FOR URBAN QUALITY

We have seen that there is usually a relationship and an interaction between urban roles, change and growth. These will of course have a knock-on effect upon urban character and quality. We have also seen that there is similarly an interaction between 'issues' and 'indicators', when we use 'environmental capacity' as a major regulating tool for urban quality. In exploring the link between issues and indicators, we should be clear what they mean. Issues are normally the situations of pressure upon the urban form or character (or deviants from the norm) which are seen as problematic/undesirable. Indicators are the outward 'tell-tales' as to the effects or impacts of these pressures. Importantly, they can arise either in the form of clues as to the nature or root of the problem(s), or in terms of an identified need for regulatory elements. We may therefore talk either in terms of indicators or regulators of environmental capacity, as affecting urban quality.

To illustrate the matter further, we shall draw again upon the Chester study. The main issues and indicators for the city's planning and urban quality problems were set out, together with much of the foregoing material on Chester's environmental capacity, in an article by Sandra Roebuck of BDP in the RTPI Journal *Planning Week* on 16 June 1994, which we gratefully acknowledge.[83] Some of these issues and indicators are listed below, the salient point being that the indicators are used to establish whether the pressures on the city comprised in the issues are increasing or easing.

City of Chester: (Some Main) Issues and Indicators

Issues for Environmental Capacity	Indicators (Clues or Regulators for Urban Quality)
• Large development in city centre impacting on historic buildings, townscape and 'urban grain'	• Change in historic building areas • Loss of critical streets, paths and alleyways over time
• Development of tall buildings impacting on the townscape, setting of historic buildings, and important views	• Visual analysis
• Under-use of historic buildings affecting their condition	• Vacancy rates in listed buildings and degree of vulnerability
• Adaptation of historic buildings to accommodate modern retail and business needs, affecting internal and external building fabric	• Condition of listed buildings • Pressures for retail presence both in town and out of town • Loss of archaeology
• Tourism impacting on comfort of pedestrians, type of shops, and the historic fabric of city	• Degree of crowding in streets • Type and range of shops • Damage to buildings
• Car commuter congestion affecting journeys and buses.	• Inbound morning peak traffic flow
• Pedestrian/vehicular conflict affecting pedestrian safety/comfort	• Conflicts between pedestrians, delivery vehicles, buses and general car traffic flows

It will be noted that all the above issues and indicators have direct planning control and environmental quality implications (some others of a purely strategic planning nature and content have been omitted). This in turn carries two further basic implications:

- Planning issues must carry primary weight and precedence in dealing with development and urban quality and design issues.
- Local planning authorities need to exercise a primary role in all such matters, both at strategic and (detailed) local level.

We have already set out earlier in this book the case for reinstatement of local authorities' central role regarding urban quality in the UK (offsetting their gradual emasculation in real terms since 1979), so there is no need to restate that theme in depth.

What we should not omit to emphasise, however, is the need to include in our environmental capacity thesis both of *the dual aspects of opportunity and restraint* (i.e. not solely the latter) since cities and planning must be able to absorb necessary change in a measured, controlled and acceptable way. It is not the intention (nor should it be the outcome), to stifle a city's natural impetus/tendency for growth. Urban quality in development however is at a premium, and techniques such as environmental capacity control have now become pre-eminent in the fight to safeguard standards of urban life.

URBAN SITUATIONS: CASEBOOK

We continue this section with commentary on a number of urban situations, via examples drawn from Europe and the UK, North America and Southern Africa, where a brief analysis or description of each of these is helpful towards a better understanding of the relevant type of situation.

Each situation is placed or grouped under a more general topic heading or 'theme', reflecting urban quality as it appears in these different contexts, via illustrations wherever possible.

Context/Capacity

Chester

We have examined already as one example, the city centre of Chester from the viewpoint of environmental capacity, in terms chiefly of the 1994 BDP/Arup study, as basically affecting urban quality, and have pointed out how the methodology used therein can be applied with benefit to other urban areas (not exclusively historic). (Fig. 20, pp. 166–7).

La Rochelle

Another interesting, but quite different example of the application of an environmental capacity approach is that of the French Atlantic port of La Rochelle. Now booming as a maritime centre for tourism, sailing, leisure and holiday accommodation, the historic inner harbour area was in danger of becoming totally swamped by the additive effects of all these activities. The town council and port authority (who had to maintain a fishing fleet as well as a yachting harbour in addition to conserving the historic character of the town generally, catering for residents as well as for visitors), came to the logical and inevitable solution of creating a 'port-plaisance' further along the coast at Les Minimes, linked to the old harbour 'core' area of La Rochelle, thus ensuring its continued vitality without detriment to it environmentally. The historic centre of the town has also benefited from pedestrian and traffic-calming measures, including public-hire bicycles (in distinctive yellow) provided by the town council for use in preference to cars – a very far-sighted move, ahead of its time (1976) in France. (Fig. 22, pp. 170–71).

Vue du Port de la Rochelle

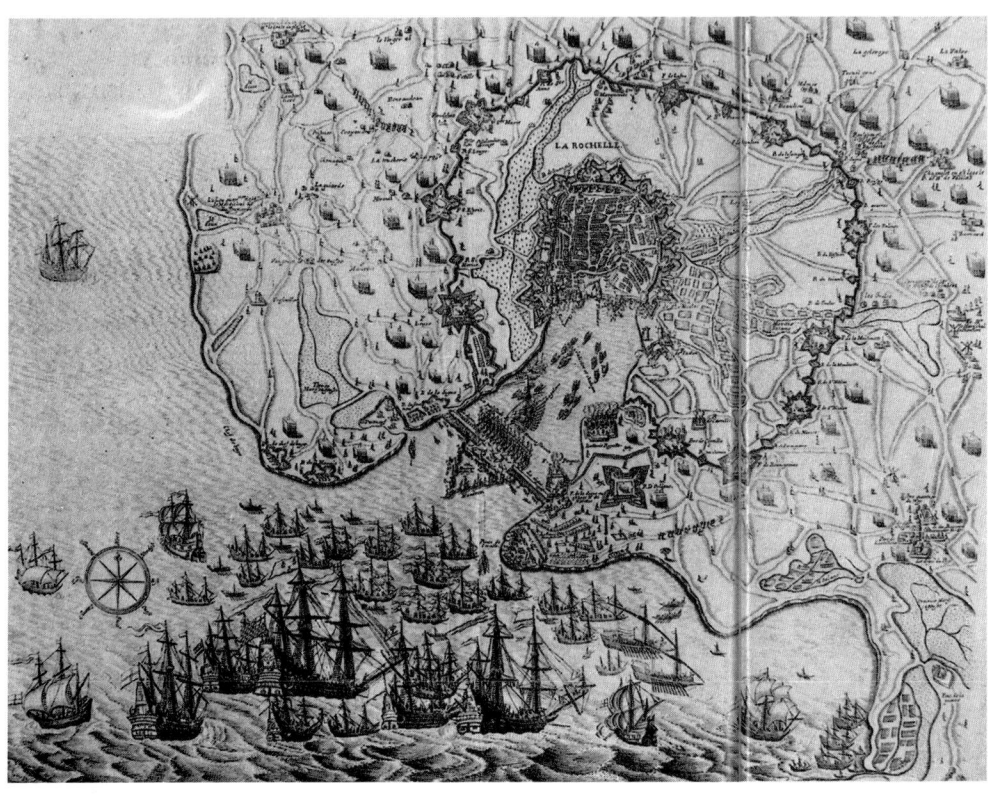

Figure 22 La Rochelle: historic inner port.
The French Atlantic port of La Rochelle is an excellent example of response to historic context and environmental capacity, in terms of adaptation to change whilst conserving its distinctive character down to its present boom in sea-related tourism and recreation. Note how little the harbourside has changed from the 1628 map and the 1749 engraving. The creation of a large outer marina and bathing beach at Les Minimes has saved the old inner harbour from saturation. The new port-plaisance is the outcome of 'environmental capacity' planning. The town centre has also benefited from traffic relief measures.
Source: *La Rochelle* guidebook, Artaud Frs, 44470 Carquefou, Nantes

Context/Conservation

Edinburgh[84]

An excellent example of a city enjoying an unrivalled natural setting, which has developed over hundreds of years into an equally eminent historic context, is that of Edinburgh. Here, the Castle, New Town, Princes Street Gardens, The Mound, Old Town, Holyrood House, and Calton Hill are in classic overall combination, through the natural topography. In the case of the Castle, its permanent visual enhancement could however be achieved by relocating the annual Military Tattoo from the Castle Esplanade (covered for long periods of the year by scaffolding to provide raked seating for the Tattoo), to a new site in Princes Street Gardens. The entire Castle, plus the rocky outcrop on which it sits, would then become an impressive total backdrop to the magnificent display of the Military Tattoo. (Fig. 17, pp. 141–3).

In the case of the 18th-century New Town, it has now been found that modern office requirements are frequently incompatible with the Classical-style buildings of Robert Adam (erected to the 1797 plan of James Craig), and return to residential is regarded as the preferred course of action in many cases.

These comments overlie a greater problem, however. The major concern of moneyed institutions, amenity groups, learned societies and the city council nowadays is with the future role of Edinburgh *as a whole* – how the city is to break free, in effect, from the cocoon of its superb topographical setting and its magnificent architectural heritage, and emerge into the 21st century embracing the future with a vision that will ensure its rightful place as the capital of Scotland, on equal terms with other European regional/national cities.

The need to innovate, in order to avoid the 'museum' image, has however been startlingly recognised in the form of Terry Farrell's 1995 Edinburgh International Conference Centre.

The new Conference Centre occupies a prominent site, readily visible from tourist viewing-points and from the principal westward artery which it adjoins, leading to the airport, yet close in to the central area of the city. On the master plan, it provides a clearly designated keynote. The Centre in fact is part of the Exchange, a 9-acre, £320m city-centre development master-planned by Farrell and seen by Edinburgh as the key to international status and consequent prosperity. Economically, it is a significant project, as it should attract annually some £19m of expenditure by visitors, and provide 640 new full-time jobs. Architecturally, the building is also noteworthy for its technical innovation. Its core is a massive drum of simple architectonic volume, within which is a 1,200–seat auditorium. Two 300–seat revolving auditoria can be incorporated into the main hall or used independently (for large events, the nearby Usher Hall with its 2,200 seats can be used in conjunction with the EICC). Finally, as a prime example of Late Modernism in architecture, the building will celebrate the 20th century, with benefit for the 21st.

Bristol

In *Bristol*, the differing claims of preservation and conservation may be examined in the context of the city's commercial heart, with its impressive heritage of buildings from medieval to Victorian times, traffic-free but now largely 'for disposal', due to the movement of economic change. Again, in Bristol, the aesthetic basis of medieval town planning was ignored when 20th-century commercial building blocked the downhill vista of St John's Church spire and St Mary le Port's tower beyond, and when pedestrianisation turned hitherto busy local trading streets into somewhat anonymous areas of gift shops and 'sanitised prettiness' (Fig. 8, p.43).

Lewes, East Sussex

As a further study of 'conservation in context' (easily seen in a day visit from London), an interesting comparison may be gained by contrasting the Castle Banks area of Lewes, East Sussex, with Sussex Square, the centrepiece of Kemp Town in Brighton. They are, respectively, excellent examples of *urbs in rure* and *rus in urbe* – i.e. the relationship between development and green open space – and they even have a historical and visual affinity in their formation and layout, albeit at different scales. The humble terraced dwellings arranged along the contours below Lewes Castle thereby enjoy enviable views and use of the green valley hollow (a short step from the town's high street); whilst the sophisticated Regency crescent of Sussex Square in Brighton has a similar, though more formalised, relationship with its own green urban space. (Fig. 10, p. 65).

Development Quality: Award Winners and Other Successes

Broadgate, London/Castle Mall, Norwich

Two recent developments of particular note for their quality of planning and design are the Broadgate scheme in London and the Castle Mall scheme in Norwich (Fig. 14, pp. 94–5). Both schemes won the Royal Town Planning Institute Silver Jubilee Cup, presented each year to developments showing an exceptional level of planning achievement – the Broadgate scheme in 1992, and the Castle Mall Shopping Centre scheme in 1994.

In the judges' view, Broadgate (adjoining the newly rehabilitated Liverpool Street Station in the City of London) 'represents a model for a high density city centre redevelopment ... which has resulted in an unusually user-friendly, well functioning, mixed-use city centre environment with optimum public transport accessibility'. In the case of Norwich, a major shopping mall was successfully integrated into the Castle hilltop site by building it underground on four levels with toplighting provided as part of a landscaped urban garden.

Amex House, Brighton

A further example of development quality, if less recent, is Amex House in Brighton, the European headquarters of American Express.[85] We make no apologies for the fact that this is not only a relatively well-established example, but is also another example from Brighton. In *A Guide to the Buildings of Brighton*:[86]

> Amex House dominates the sweep of Carlton Hill. Built in 1977 to designs by Gollins Melvin and Ward, it makes a much more successful addition to Edward Street than its contemporary neighbours. The horizontality of the main elevations fails to disguise the building's bulk, but the architects avoided adding another tower block to Brighton's skyline. Indeed, the combination of white GRP panels, used here for the first time in this country and still impressively clean, and blue tinted glass, seems to be an appropriate one for a contemporary Brighton building.

The American Express offices, though dating from the 1970s, are one of the best modern buildings in Brighton. Not only of a classical, pagoda-like elegance in mass and detail, the building enhances a wide area of urban background generally. It pulls together a fractured tableau of banal tower blocks, provides a visual focus for the area between the Steine and Kemp Town, and makes an interesting comparison with other buildings of individual character such as the Dome and the Royal Pavilion. Unloved by preservationists, it is nevertheless an example of the need to consider the wider context, as well as the closer/more innovatory aspects or impressions of any building. (Fig. 11, p. 67).

Public Realm: Public/Private Sectors

Metropolitan/macro scale

Planning and building in the 1950s and beyond in major cities of both Britain and Europe, as well as in the Americas, was characterised by high-rise development almost everywhere, either directly carried out or promoted by city authorities, often following Le Corbusier's impersonal *Ville Radieuse* planning ethic, ignoring social needs. The Loughborough Road housing estate in London has had a strong influence of this kind on public sector design both then and since.

In Britain, an intelligent approach to providing a large amount of mixed-use development with a cultural content was London's Barbican scheme in the heart of the City, albeit still at macro scale. In America, the revitalisation plan for Fort Worth, Texas, with its traffic/pedestrian provisions, was echoed later in the 1960s by the 'ring and loop' road plan for the city centre of Norwich. (Fig. 18, pp. 162–3).

Local/humanist scale

Following the demise of most forms of public sector development in Britain, due to its political/economic policy, increased attention must be given to any opportunities for joint public and private sector development on a local/humanist scale. An excellent example of this is LDR's scheme[87] for the enhancement and future growth of Greenville, South Carolina, USA. This far-sighted scheme provides for a very attractive mixed-use development at Riverplace, of which a major imaginative feature is a semi-circular central lawn both relating to and continuing the amphitheatre of a Performing Arts Center across the river. (Fig. 19, p. 164).

It is worth remembering that the concept of a central cultural/arts element contained in a public space by a formal enclosure of residential and other buildings was anticipated many years before by Alberti in Italy, and later in the 19th-century layout for the area around the Royal Albert Hall in London. History is still relevant.

Public Realm: Midlands/North of England/Scotland

Birmingham: Victoria Square

The revival of Britain's premier Midland city is superbly demonstrated by the civic renovation of Victoria Square, now a major traffic-free focal point in the city centre, located on a pedestrian axis.[88] Full advantage has been taken of the slope across the site plus the major civic buildings on two sides to make a scheme of the highest quality, based on an upper and a lower precinct, linked by a grand assemblage of steps, sculpture and water. The scheme stems from the Highbury Initiative of 1988, which saw the need to transfer most of the traffic from the Inner Ring Road to Middleway, upgrading the latter and thus facilitating the pedestrianisation of Victoria Square and its civic design renewal. (Fig. 23, pp. 176–7).

Victoria Square incorporates (*inter alia*) Birmingham's old Town Hall in the context of the upper and lower precincts. The purity of its Roman temple design is cleverly reflected in the many historical references embodied (in a modern way) in the sculpture and layout of the new civic square. In the layout, there may perhaps be distant references also to Michelangelo's design for the Capitol in Rome: but Victoria Square welcomes rather than overwhelms us, while the fine quality, durable materials used here still speak of civic pride and affirm the importance of a user-friendly public realm. (Fig. 24, p. 178).

Manchester: GMEX Area

In the North of England, Manchester's GMEX Area has the potential to be one of the most exciting and comprehensive visitor magnets in Britain. A concentration of existing/proposed facilities and cultural attractions, and the new Light Railway system, with the old Central Station converted to an Exhibition and Event Centre and situated at the core, together provide a high-quality urban complex and an excellent example of the value and wisdom of investing in the public realm. (Fig. 25, pp. 180–81).

Manchester: Light Railway

From its pioneering days of creating the original 'trams' as an urban public transportation system, Manchester has once again proved a brand leader – its new light railway provides effective and attractive means of mass transit, geared both to the needs of the public and the context of the urban environment. The Light Railway connects the city centre with outlying areas, utilising old 'tramway' track and suburban rail track, and serving *inter alia* the GMEX Area, with its developing conference, cultural and civic facilities described above. (Fig. 26, p. 182).

Manchester's Light Railway served also as the forerunner for the *Sheffield Supertram* system,[89] due to be fully operational in autumn 1995, after phase 1 opened in 1994. The scheme cost £240m (mainly government funded) and is the largest public transport project to be commissioned in the last 20 years outside of London, since the construction of the Tyne and Wear Metro. It is of particular relevance to our thesis that the system was constructed to an urban design brief, produced at the outset of the scheme by the local authority (the city council) marrying together both engineering and environmental/physical appearance considerations. More fundamentally, Light Rail Transit systems such as in Manchester and Sheffield are, according to some, the only viable form of mass urban transit for the future. The LTR comes in a user-friendly form, quiet and pollution-free, and with an impeccable policy pedigree as an essential ingredient in achieving Local Agenda 21, in reducing the need for petroleum and hopefully reducing the need for private transport. This is a significant step in effecting a modal shift away from the private vehicle (in line with PPG13), and in turn helping to reverse the pernicious draining away of activity (particularly shopping) from the city central area to the outskirts.

Manchester: civic scale

Manchester has inherited a rich legacy of civic-scale buildings from the 19th century, which still provide the architectural and environmental context for the city centre. A few examples, taken almost at random, demonstrate this. The old Royal Theatre and the Midland Hotel, both effectively within the GMEX Area, have this civic character and scale (even the new office block next to the hotel reflects the same scale); while the circular Central Library and the commanding Town Hall block (built in 1938 as an extension to the Victorian Town Hall) together dominate St Peter's Square (Fig. 27, p. 183). Unfortunately, the low-key insubstantial 1960s style block, on the Oxford Street corner of the square, is a poor neighbour to the civic buildings opposite – a commentary on prevailing economic considerations and lack of visual/environmental perceptiveness at the time. However, one only has to turn the corner into Albert Square to experience a more happy marriage of Victorian splendour with modern civic scale.

Glasgow: historical context[90]

> Glasgow is the city of business, with the face of foreign as well as of domestic trade. It is a large well-built stately city standing on a plain in a manner four-square, and the four principal streets are the fairest for breadth and the finest built that I have seen.
>
> (Daniel Defoe, 1704)

Nearly 300 years and an Industrial Revolution later, Glasgow's selection as City of Architecture and Design 1999 provides a relevant theme to the following examples of the city's

Figure 23 Birmingham: Victoria Square.
The revival of Britain's premier Midland city is superbly demonstrated by the civic renovation of Victoria Square, now a major traffic-free focal point in the city centre, located on a main pedestrian axis (above and opposite).
Sources: Photo: Michael Parfect; Plan and line drawing: *Urban Design* Journal, 54, April 1995 (drawing by Francis Tibbalds, planning design consultant)

urban quality. The first group relates to its historic character. First, the Italian Centre in the city centre is a mixed retail/restaurant/residential scheme forming a prestigious focal point for the well-known Merchant City redevelopment of 1980s origins. Second, the Atlantic Quay offices (phase 1) on the north bank of the Clyde exemplify 'Glasgow style' modern architecture (note the careful relationship with the Clyde Port Authority Building). Third, Templeton's Carpet Factory (now a small business centre) shows the Victorians' flair for disguising an 'industrial shed' in Venetian Gothic facade detailing. Fourth, St Andrew's Church (17th century), now an arts centre, has a new neighbour: a neo-Georgian residential terrace, providing a high-quality street environment albeit at high unit cost somewhat at odds with a rather run-down part of the city.

Finally, as a further example of Glasgow's contribution to urban quality, we may cite Princes Square, where a soaring curved tubular steel and glass roof above Victorian facades creates a speciality shopping court reminiscent of the 'Golden Age of Glasgow' and a magnet for many visitors worldwide. (Fig. 28, p. 183).

PERSPECTIVE ON URBAN QUALITY 177

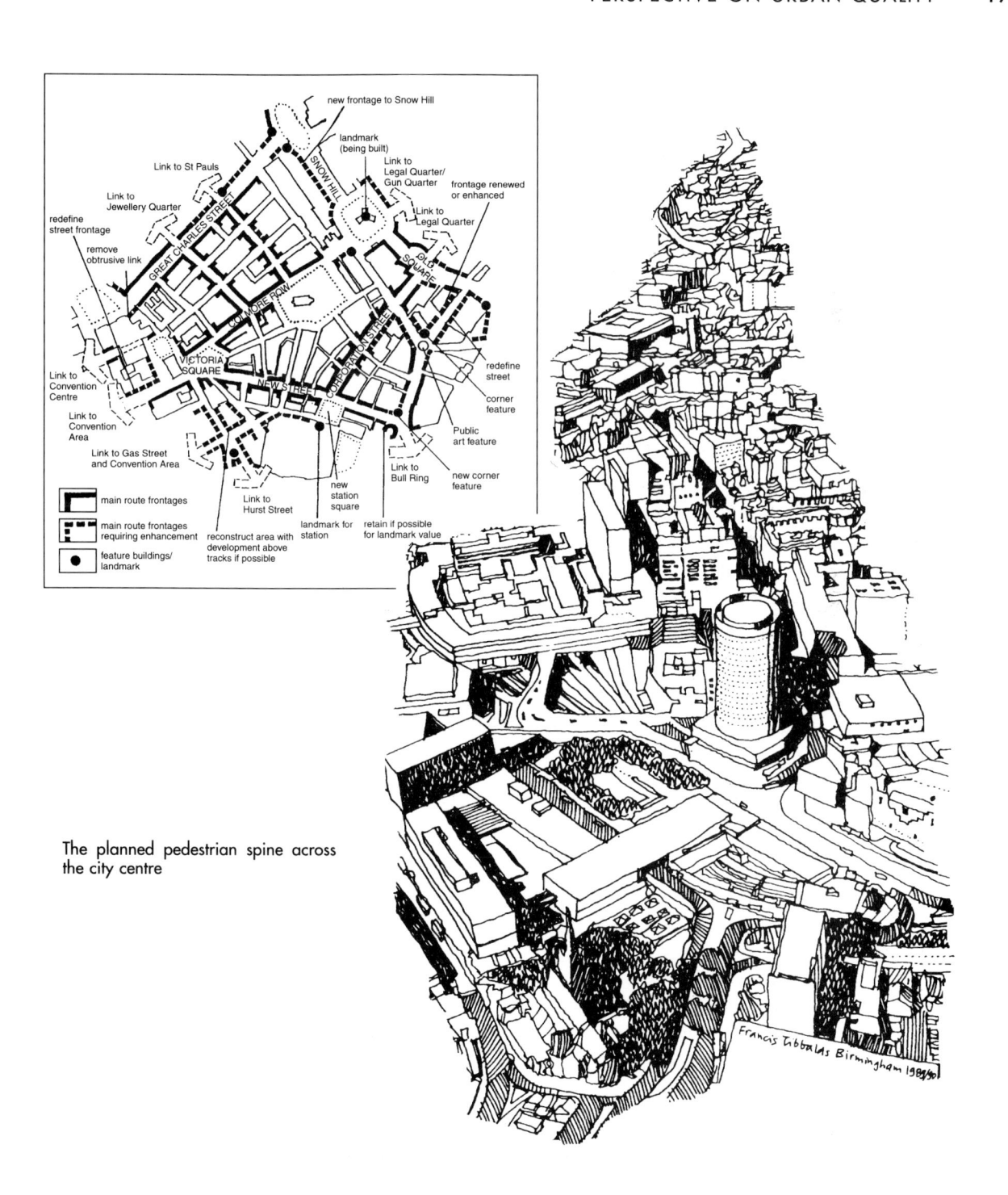

The planned pedestrian spine across the city centre

178 PLANNING FOR URBAN QUALITY

Figure 24 Birmingham: Town Hall.
This view of Victoria Square shows Birmingham's old Town Hall in the context of the upper precinct. The purity of its Roman temple design is cleverly reflected in the many historical references embodied (in a modern way) in the sculpture and layout of the new civic space. In the layout, there may perhaps be distant references to Michelangelo's design for the Capitol in Rome but Victoria Square welcomes rather than overwhelms us, while the fine quality and durable materials used here still speak of civic pride and affirm the importance of a user-friendly public realm.
Source: Michael Parfect

Glasgow: modern tenements

Glasgow's return to traditional scales and forms via the advent of the modern tenement block is particularly welcome in a city badly scarred by postwar slum clearance and mass social housing projects, with their legacy of crumbling peripheral estates and unloved high-rise blocks. As examples, the Byres Road scheme near the university blends easily with the Victorian buildings adjacent via sensitive design (and stone cladding, financed by Glasgow Development Agency). Second, the St George's Cross scheme incorporates an elderly persons' residential home, complete with dining/social area. Additionally to this, the Crown Street Regeneration Project, at work in the Gorbals, has produced a mixed-tenure jointly funded scheme replacing a faceless high-rise block, and introducing a human scale of development to help cure the area's endemic social problems.

Public Realm: Waterfronts

The reflective air and reposeful quality of a major body of water ties all together in a symmetry about the plane of its surface, and its application or existence in an urban context has been the making of many town/cityscapes throughout history. It is no accident that London, Bristol, Liverpool, Cardiff and many other of our old commercial ports have found high residential land values along their deserted docksides. An increasing consideration of significant 'waterscape' through our urban planning and architecture should be utilised in new urban development generally.

The two northern waterfront cities of *Newcastle* and *Liverpool* (Fig. 29, pp. 184–5) repay attention/study from an urban quality point of view. Both share the same sort of debilitating problems, yet both have similar potential for enhancement/revival, given the right economic circumstances or approach. Views of Newcastle's Tyne Bridge show the possibility for maximisation of a prime waterfront position in both commercial and aesthetic terms, while exploratory design studies[91] for Liverpool's Chavasse Park provided for mixed uses around a central civic route, relating with the Pierhead and the Mersey.

A city with a great maritime trading past, *Bristol* must now look to the future, whilst conserving and making best use of its superb natural setting and its splendid architectural legacy of 200 years or more. Despite its busy commercial life, an air of uncertainty seems to lie over its quaysides and historic precincts, where trade and commerce once thrived. New uses are needed increasingly for open waterfronts and neglected old buildings if Bristol is to remain in touch with and true to its distinguished past, and if urban quality is not to be forgotten. (Fig. 16, pp. 138–9).

Another key issue to be considered in the overall subject of waterfronts is the relationship between *bridges and buildings.* Too often we think of them separately but the two elements must be in balance in both visual and functional terms. By way of example, the bridge over the River Scheldt in Ghent, Belgium, with its impressive group of fine stone commercial buildings is comparable to *York's Lendal Bridge*, with the 1994 General Accident Life Headquarters building, which has been acclaimed for its qualities of design relative to its riverfront setting.[92] (Fig. 30, p. 186).

The New South Africa: Cape Town

The ending of apartheid, under Nelson Mandela, has opened up a new era in South Africa. Accordingly, the country is opening up to an unprecedented influx of tourists and other visitors, business people, etc. from Britain and Europe. Commensurate with this is a renewed interest in what South Africa has to show, in terms of quality of urban life and development, as well as its undoubted natural advantages.

Without wishing to gloss over the past evils

180 PLANNING FOR URBAN QUALITY

Figure 25 Manchester: GMEX Centre.
Manchester's GMEX Area has the potential to be one of the most exciting and comprehensive visitor magnets in Britain.
Sources: photo of GMEX Centre building: Gordon Power; aerial photo of GMEX District/development layout for GMEX Area: North West Tourist Board/Land Design Research International (LDR) – Manchester, Salford and Trafford Tourism Development Initiative, January 1993

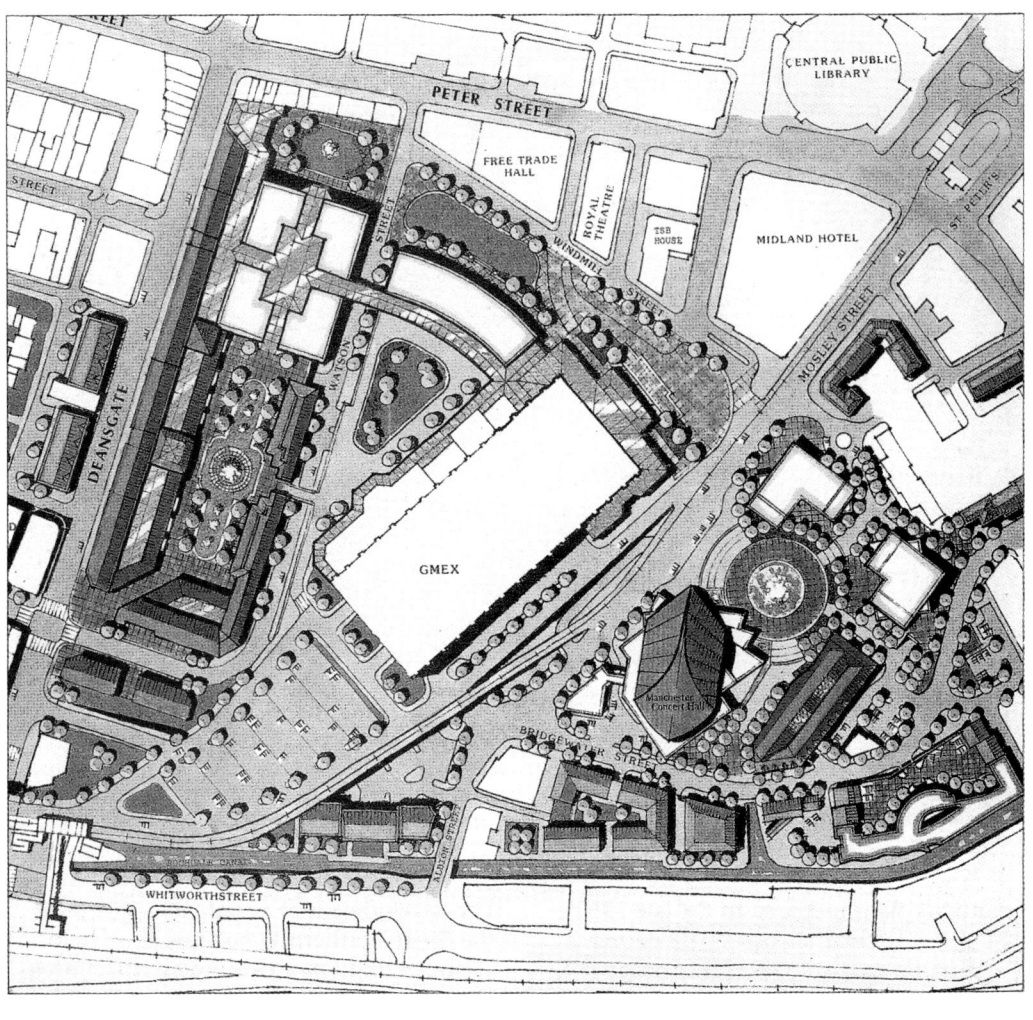

of apartheid and the legacy it has left, which will take many years to deal with, we should examine with an impartial eye the very many achievements in the field of urban quality in South Africa, and give full credit where it is due – especially where we can learn lessons ourselves from these achievements.[93]

Lessons from Cape Town

The city of Cape Town enjoys an unrivalled natural setting at the foot of Table Mountain and along the sweep of Table Bay. It can be clearly seen how the Central Business District has formed close to the docklands area, and how

Figure 26 Manchester: Light Railway.
From its pioneering days of creating the original 'trams' as an urban public transportation system, Manchester has once again proved a brand leader with its new Light Railway (shown here in operation in the city centre).
Source: Michael Parfect

development has followed the main natural features/contours. It can also be seen that planning control, via zoning, has very largely kept high-rise commercial/civic buildings confined within a tight central area – the classic/ideal urban profile. The city centre however will need to compete with the magnetism of the revitalised Waterfront area, booming from new tourism and shopping. (Fig. 31, p. 187).

The Victoria and Alfred Waterfront development visited in Cape Town is both a highly successful commercial venture and an object lesson in the revitalisation of a working dockland area which retains and enhances historic buildings with the addition of equally attractive new ones – all to a uniformly high architectural/environmental design level. This remarkable development complex, which includes many shopping, leisure, crafts, restaurant, accommodation and service facilities, has been so successful that it has competed unexpectedly forcefully with commercial activity in the city centre. Further extensions are planned to the Waterfront, including a marina and residential units, due to open in 1996/97. (Fig. 32, p. 188).

Some examples of *public and private sector housing* in Cape Town, also visited in the study tour, contrast interestingly. Each represents in its own terms a desirable level of environment and at the same time a departure from the

PERSPECTIVE ON URBAN QUALITY 183

Figure 27 Manchester: Civic Scale.
Manchester has inherited a rich legacy of civic-scale buildings from the 19th century, which provide the architectural and environmental context for the city centre. The circular Central Library and the commanding Town Hall block (built in 1938 as an extension to the Victorian town hall) dominate St Peter's Square.
Source: Michael Parfect

Figure 28 Glasgow: Historical context, new life.
Glasgow's selection as City of Architecture and Design 1999 testifies to the city's urban quality. The Italian Centre in the city centre (left) is a mixed retail/restaurant/residential scheme forming a prestigious focal point for the well-known Merchant City redevelopment of 1980s origins. The Atlantic Quay offices (phase 1) (right) on the north bank of the Clyde exemplify 'Glasgow style' modern architecture (note the careful relationship with the Clyde Port Authority Building).
Source: George Benson

Figure 29 North of England waterfronts: Newcastle/Liverpool.
The two northern waterfront cities of Newcastle and Liverpool repay attention/study from an urban quality point of view. Both share the same sort of debilitating problems, yet both have similar potential for enhancement/revival, given the right economic circumstances or approach. The view of Newcastle's Tyne Bridge (above) shows the potential for maximisation of a prime waterfront position in both commercial and aesthetic terms, while the exploratory design studies for Liverpool's Chavasse Park (opposite), providing a mixed-use scheme around a central civic route, relate with the Pierhead and the Mersey.

PERSPECTIVE ON URBAN QUALITY   185

Sources: Newcastle photo: V.K. Guy Ltd, Windermere, Cumbria; Liverpool: Land Design Research International (LDR), Columbia, Maryland, USA

LIVERPOOL
Chavasse Park

186 PLANNING FOR URBAN QUALITY

Figure 30 Waterfronts: bridges and buildings. The bridge over the River Scheldt in Ghent, Belgium (right), with its impressive group of fine stone commercial buildings is directly comparable to York's Lendal Bridge (above), with the 1994 General Accident Life building, which has been acclaimed widely for its qualities of design relative to its riverfront setting.
Sources: GA Life building: *Architecture Today* no. 45, Feb 1994 (photo, Jeremy Cockayne, St Paul's Square, York); Ghent photo: M. Parfect

Figure 31 Cape Town: context and scale.
Cape Town enjoys an unrivalled natural setting at the foot of Table Mountain and along the sweep of Table Bay. We can see how the central business district has formed close to the docklands area, and how development has followed the main natural features/contours.
Source: Michael Parfect

188 PLANNING FOR URBAN QUALITY

Figure 32 Cape Town: Waterfront vitality and success.
The Victoria & Alfred Waterfront development in Cape Town is both a highly successful commercial venture and an object lesson in the revitalisation of a working dockland area.
Source: Michael Parfect

norm. On the one hand, Springfield Terrace, Woodstock, a pilot project carried out by a housing association jointly with the city planning authority, is resettlement housing for coloured people – in sharp contrast to the acres of degraded shantyland in which some black people live. On the other hand, *the Marina da Gama* is a planned waterfront development of middle- to high-income housing, situated on the Muizenberg estuary south of Cape Town. It is an area of what some might call 'elitist' or even 'segregationalist' development – it tends to turn its back on the outside world physically – but in layout planning and design terms, and in overall visual appearance especially when seen in the wider landscape context, it scores extremely highly.

In terms of specific provision for black/coloured areas, major problems still exist as to the continuing presence of unconscionable areas of shantyland, which spread all over the Cape Flats, and which suffer the related problems of high winds and flooding. The city authorities have attempted to alleviate the problems of some of these areas by developing shopping complexes at Guguletu and Nyanga Junction. The latter is a large, multi-level modern complex, integrated with the railway station of that name, and providing a high standard of shopping and trading facility. It is also the first major black-managed development in South Africa. In spite of this, there is a prevailing view amongst the black population that they would have preferred a number of smaller, more localised, neighbourhood shopping centres, and associated social and care facilities, with which they could identify and which would each serve as a nucleus for the housing area grouped round it.

In the words of Cliff Hague,[94] 'Cape Town's black townships are monuments to apartheid planning – the government instructed the local authorities on the layouts'. Thus, the new town of Khayelitsa was created (bulldozing the shacks, but still far from employment areas).

Reference should also be made to Cape Town's *District 6*. This was laid out in the 1830s, on the lower slopes of Table Mountain, was settled by former slaves, poor whites, and indigenous peoples, and became 'a vibrant, multi-ethnic cultural melting pot – negating ideologies of racial purity'. It was recently cleared, despite the scale of the opposition aroused. 'The solution found for the area was to let the market develop the vacant land for an extension of the city centre. The value generated could then be used to recompense previous residents and also boost funds for meeting the housing needs of all the city's poor. But a planning solution would have been to build on the district's history, assets and location'.

It must now be seen whether strategic planning, economic development and participation will replace detailed top-down restrictive planning in the new South Africa.

South-West Africa: Windhoek, Namibia

Windhoek, the capital city of Namibia (formerly German South-West Africa) is a 'city in transition'. The development plan for its Central Business District shows, in its Post Street tree-lined, open-air, pedestrian mall proposals, that the European lessons of car and pedestrian conflict have been learned, even at a time and place where the worst excesses have yet to inflict themselves. *Rus in urbe* is still the intention. Fountains, sculptures and monuments are, with greenery, seen as important in a city centre which otherwise might have become a car-dominated assembly of functional commercial blocks. The need for civic pride and urban quality has been fully recognised, in which people come first. Windhoek retains its roots and reflects its Afro-European character whilst incorporating the European experience of urban design in moving into the 21st century.[95]

Figure 33 Windhoek, Namibia: A City in Transition.
The European lessons of car and pedestrian conflict have been learned, even at a time and place where their worst excesses have yet to inflict themselves. *Rus in urbe* is still the intention. Fountains, sculptures and monuments are important. People come first (above and opposite).
Source: Barrie P. Watson, Chief Town Planner, City Engineer's Dept, Windhoek, Namibia

Design and Technology

New and old

Architectural design must relate equally to purpose and context. It must also make use of materials and construction best suited to both. All these factors, combined with the personal qualities of design skills and sensitivity, are needed for the result to be appropriate and visually pleasing. New design has something to learn from older buildings in this; even if mass and scale are different, problems and approaches may still be necessarily similar.

Modern design and construction

We do not propose to enter upon a comprehensive tour or thesis concerning modern architecture – this is far too wide a subject for inclusion as a mere section in a book on urban quality generally, and in any case has already been amply covered elsewhere, via professional journals and literature.[96] All too often, however, modern buildings are seen mistakenly by the layman in terms of style – whereas the predominant factor now is that of *new technology*. The future, in terms of architecture, will increasingly be shaped by developments in technology and also in standards of building construction, as much as by urban design. (Fig. 34, pp. 192–3).

In this, the high quality of design and construction for which modern Danish[97] and German buildings are renowned should be noted. Developments such as the technologically advanced, multi-storey car park (1991) at Copenhagen International Airport, or in the field of prefabricated, low-rise housing or office headquarters buildings, provide good examples

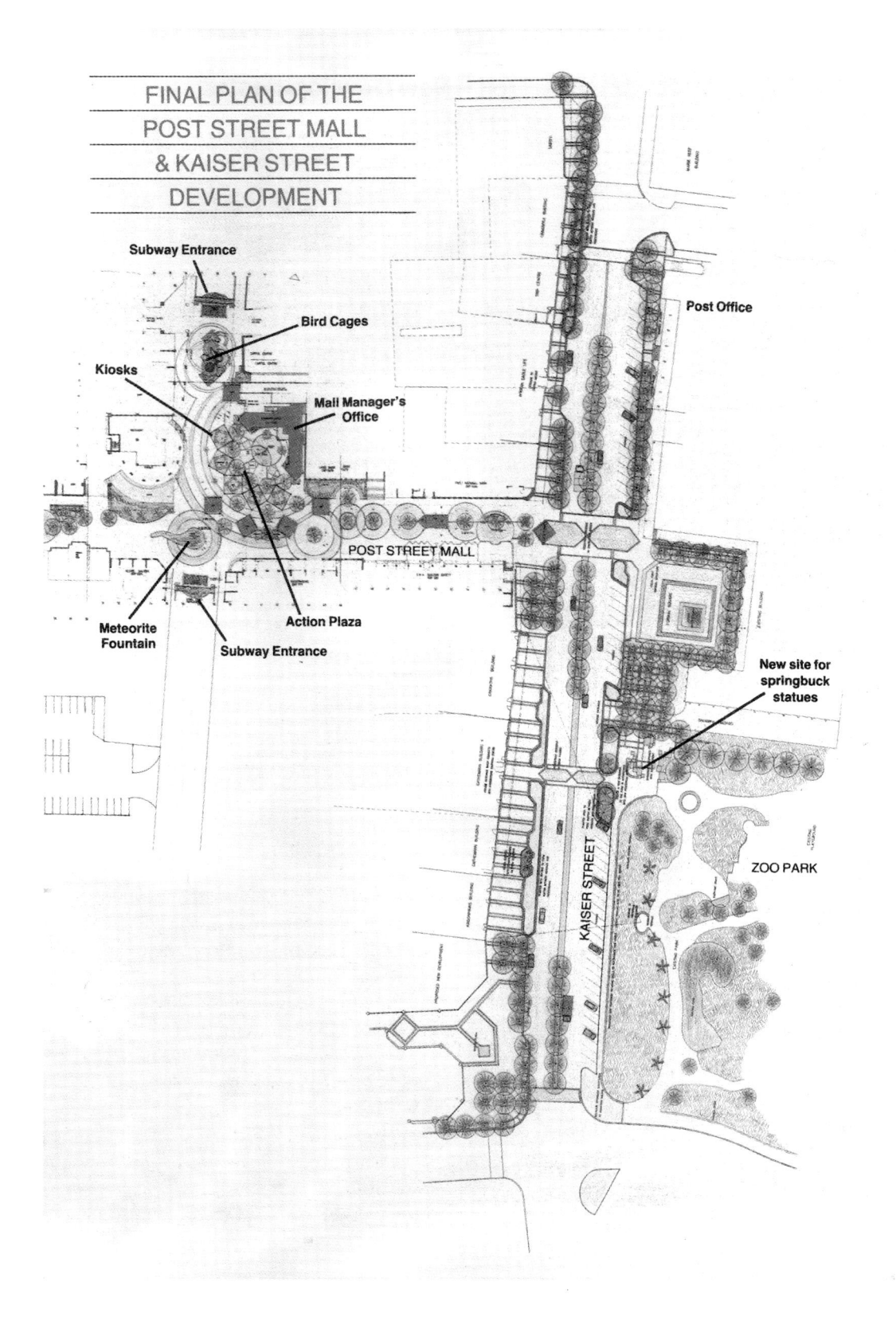

FINAL PLAN OF THE POST STREET MALL & KAISER STREET DEVELOPMENT

- Subway Entrance
- Bird Cages
- Kiosks
- Mall Manager's Office
- Meteorite Fountain
- Action Plaza
- Subway Entrance
- POST STREET MALL
- Post Office
- New site for springbuck statues
- KAISER STREET
- ZOO PARK

Figure 34 Danish building excellence: recent examples.
The high quality of design and high standard of construction are exemplified in these three illustrations of modern structural technology, all from the same firm of civil engineering consultants and contractors in Denmark.
  The illustration above shows a multi-storey car park at Copenhagen International Airport, completed in 1991. The illustrations opposite show the application of industrialised building techniques to some recent Danish commercial office and housing schemes.
Source: Højgaard & Schultz, Jaegersborg Alle 4, DK2920, Charlottenlund, Denmark (Colin Gosden)

of the quality that is achievable today in modern industrialised buildings. Indeed, engineering technology and engineering solutions to building projects are increasingly to be found in the modern world. The 'ground-breaking' buildings of Richard Rogers (Lloyd's offices in London and the Pompidou Centre in Paris), of Norman Foster (Hong Kong and Shanghai Bank building and Stanstead Airport), and Terry Farrell (Charing Cross Station and the Edinburgh International Conference Centre) are but a few examples of the innovatory contribution which British architects are making to the international scene.

A further example is the highly imaginative building designed by the British architect Will Alsop at Marseilles, France, for the headquarters of the local authority. Alsop's *Hôtel du Département* (Fig. 35, pp. 194–5) shows the dramatic possibilities offered by modern structural methods and materials for the creation of breathtaking new architectural forms. The design draws upon both airport and engineering experience, influenced to some extent by Corbusier's Unité d'Habitation nearby, in terms of multi-purpose accommodation, but in a much lighter, celebratory mode. The aerofoil-shaped profiles of the repetitive tubular steel

PERSPECTIVE ON URBAN QUALITY   193

194 PLANNING FOR URBAN QUALITY

Figure 35 Hôtel du Département, Marseilles: modern design and technology.
The dramatic possibilities offered by modern architectural design and structural methods/materials are shown in this French-built/British designed (Will Alsop, architect) local government HQ complex in Marseilles (above and opposite).
Source: RIBA Journal April 1994

cladding to the *deliberatif*, echoed by the presidential element above, with the whole frontage assembly riding on concrete piloti above the Metro beneath, and connected back to the main office block by inclined glazed links, create a powerful and startling composition in steel and glass which commands the otherwise nondescript context of suburban villas and urban motorways.

New technology applications

An increasing role will be played by new technology in the future as to the design, control and implementation of the environment. Computers are already used to produce design/structural detailing in architecture and building. Now, computer-based visual representation in planning control can cut out lengthy negotiations and paperwork on anything from major developments to house extensions. Communications buildings and towers are now prominent in the urban scene and need to be treated as architecture, while 'remote sensing' from the air and space yields vital information in monitoring and control of traffic movement, and in planning for people and their environmental needs plus related building patterns. (Fig. 36, p. 196).

MODERN ARCHITECTURE: DESIGN AND TECHNOLOGY

The Design Process: Art, Science or Market Exercise?

A basic understanding of the design and development process is needed before a discussion of architecture, whether past or present, with its technological context, can usefully take place. This must include a basic appreciation (or recognition at least) of matters such as site selection (where any option exists); site planning and floorspace quantum requirements; vehicular and pedestrian access needs; functional and technological considerations; building and other development costs; and required level of economic return against life/use expectancy for the development. All these, plus the vital matter of aesthetic concept, must be set within the relevant urban context, including both planning restraints and urban quality expectations and demands. Architectural form and appearance is very much an outcome of these many different influences – a fact rarely, if at all, appreciated by the critical layman or preservationist, particularly when faced with buildings of Modernist functional design.

Lack of understanding of the design process, with resultant incomplete arguments about the 'design ethic' in architecture and building, has led to what can undoubtedly be considered 'The Great Divide' – old versus new – fuelled and exacerbated by ingrained attitudes and by public reaction against development pressures of all types. It is clear that in many ways this dichotomy is the ongoing preoccupation of most people in Britain with any interest in planning and development, however indirect. This theme

196 PLANNING FOR URBAN QUALITY

Figure 36 New technological applications in planning.
An increasing role will be played by new technology in the future as to the design, control and implementation of the environment. Computers are already used to produce design detailing in architecture and control of building (above), while 'remote sensing' from the air and satellite (below) yields vital planning information.
Sources: RTPI Journal, 16.3.1995/Anglia Poly. University of Chelmsford, Essex (Tony Hall); Postcard view of Isle of Wight from space, *Daily Telegraph* colour library (Clare Power)

often thus takes precedence over, or runs in parallel to, more fundamental considerations.

We shall examine this problem and its manifestations later. First, a distinction should be drawn between traditional and present-day processes and contexts.

Traditional design process/context

In past times, certainly up to the Industrial Revolution in Britain, the prevailing context for building design was one of local needs and resources, local materials, local building methods, and local briefs geared not only to the local site owner/developer and to local people but also to the availability of a competent/suitable local designer. The only real 'constraints' on development, apart from the basic ones of land and finance, were the above *local factors*, all of which could be said to be 'sympathetic' to the task, especially where a benign form of patronage was present, as with the great (and lesser) aristocratic landowners, or successful merchants, for example.

This process extended well into and in some cases beyond the 19th century, until World War I put an end to the hegemony of the land-owning gentry, in favour of the rising class of industrialists (both large and small) and the vastly growing middle classes. Throughout past times, however, from the Middle Ages to the 18th century and onwards, local factors virtually dictated the form and design of development, with one important exception: that of (imposed) architectural style. In Britain, as in mainland Europe, the spread of building styles, where not purely 'vernacular', were largely influenced by or sprang directly from Classical architecture. This *lingua franca*, or common architectural language, held good for centuries, and has not entirely disappeared yet, as can be seen in many a Neoclassical 'pastiche' office or residential development, admired by many, but representing the distorted echoes of a vanished world today.

Modern design context and process

At the time of the 19th-century Industrial Revolution, not only in Britain but also in Europe and in the Americas (indeed, throughout the Western world), the arrival of new technology in the form of steel frames for large structures and spans, and also sheet glass, went hand in hand with greatly increased affluence, mobility and space requirements. All of a sudden, the new social and economic situation produced building needs, methods, materials and resources that were *external rather than local*, to a large degree. This is the process which has been continuing right up to the present day, and will continue into the foreseeable future. The advent of unrestrained market forces and the philosophy that goes with it explains not only the huge diversity of building requirements, but also the 'ubiquity' of building materials and styles, and frequently their non-conformity in the urban scene – a headache for Planning Authorities in terms of controlling development, and dismay for a public more attuned to the cohesive architectural codes of the past.

We should not, however, omit to consider the acute difficulties faced by architects today. The 'inverted pyramid' depicting some of the modern-day hurdles encountered by building designers (reproduced with permission from the RSA *Journal* of July 1994) illustrates these difficulties well (see below). And when we view these restraints against the more fundamental constraint of the UK economy, weakened by world-wide recession, central government whims and vagaries, and the unwillingness of an inherently conservative British public to embrace fully the advantages of technological innovations in building design and construction, we need not wonder at some of the more overtly challenging attitudes displayed by architects today. Thus, when the agenda seems mainly concerned with questions of 'style', we should perhaps consider these problems more closely – as exemplified below.

. International Consortiums ... Lawyers .
. Accountants. Tax Officials .. Status .
. Politicians. Propaganda. Ethics .
. Public Relations .. Misdirection .
. Technocratic. Bureaucratic .
. Languages. Competitions .
. Bankers. Profitability .
. committee Decisions .
. Delaying Tactics .
. The Status Quo .
. Priggishness .
. No-go Area .
. Morality .
. Play it .
. Safe .
?
DESIGN

Little more need really be said, but the RSA diagram is aptly subtitled 'Without vision the people perish'! (James Gardner, RDI, designer.)

Today's scenario: market exercise

This question now pervades the whole field of building design. In broad terms, it might be said that (a) the period up to the 19th century represented (in both its vernacular and Classical forms) the time when architecture was an 'art' – i.e. a series of 'artistic essays' in building, carried out via an inherited architectural language and a restricted 'palette' of local elements and constraints; that (b) the 19th and early-to-mid 20th centuries represented the period of change when 'old' mixed with 'new', and *when 'art' mixed with 'science'*; and that (c) from then onwards the scenario is increasingly that of architecture and planning as *'market exercise'* (as we have been at pains to point out in preceding chapters), whether in old or new forms.

'The Great Divide': Old v. New

We return to the theme of this dichotomy, perceived by public and professionals alike, and examine both these aspects. The dichotomy between old and new is real enough, especially in Britain, but is likely to be replaced (be it ever so gradually) by an increasing acceptance of *'pluralism'* as the ethic of the future in the architectural and urban design landscape. This seems to be inescapably so, having regard to the *force majeure* of changing socio-economic factors on a wide scale, but the process is almost bound to be long drawn-out and difficult.

What are the main preoccupations of the participants in 'The Great Divide'? There is probably no single factor that can be identified as the one cause of the 'split', but rather there is a combination of factors that preoccupy people and perpetuate the problem. The three most cogent factors can, *at the 'human' or personal level,* be described as:

- the shock of the new;
- focus on 'style'; and
- fixation with the past.

There are, of course, other pervading factors, such as:

- educational differences;
- social class differences; and
- political colouring or outlook.

To these we might also add 'royal connotations' and public perceptions of urban quality as synonymous with preservation or conservation in Britain today. Either separately, or in combination, these result in the opposing forces of *resistance to change versus acceptance of the need for progress and innovation.*

At the 'macro' level, however, we have such factors as:

- national economic and social change and
- technological development and change (linked with building economics).

These powerful, and to a large extent overriding, considerations create a further form of tension in the situation, whereby change becomes inevitable, even if resisted, influenced or modified by all the preceding 'human' or personal predilections (for such they are). 'Macro' factors must in the end prevail, but it is perhaps both necessary and democratic that there should be a process of debate along the way, however tedious this may be, provided that neither a coarsened 'market force' credo is allowed to trample over all other considerations, nor blind prejudice allowed to undercut and sabotage all attempts at real progress. We should pause to examine the various factors we have listed as contributing to 'The Great Divide', in order to bring some constructive argument to bear on the debate.

First, 'the shock of the new'. Throughout history, the introduction of unfamiliar forms and elements in the urban scene has frequently met with protest, opposition and even derision. This applies not only to the urban scene, but also to art and music in particular – indeed, a major and seminal book called *The Shock of the New*, by Robert Hughes, has in our time examined the whole broad canvas of change in the artistic and cultural scene, to good effect. Perhaps the one message that can be drawn from an examination of 'unwelcome change' in the urban and cultural scene is that the reaction to it (though not unexpected) is most usually the result of *misunderstanding and lack of appreciation of the underlying factors that bring about such change*. Our dislike of change (almost any change) is also related to our tendency to nostalgia, and our reluctance to respond to new forms of design in a favourable way, where exercise of our intellect and our emotions is involved (we are very much victims of habit).

Second, 'focus on style'. Preoccupation with outmoded questions of 'style' still pervade the view/assessment of many a building project today. Except in specifically historic areas, where to depart widely or radically from established architectural forms could not be condoned, the concept of 'style' as being of overriding importance cannot now be valid. *'Style not style'* is the real consideration – i.e. 'style' in the sense of a display of architectural flair and design merit (even if 'new' in form), as against blind adherence to preferred or inherited 'styles' of building, either via 'blanket' preservation/conservation, or debased regurgitation of past styles in the form of 'pastiche'. Curiously, although we set great store by attractive and innovative design in things like cars and fashion, we appear to have a major 'blind spot' where design of environment is concerned.

Third, 'fixation with the past' is a causative factor closely bound up with resistance to change and blind adherence to established (historical) styles of building. Fixation with past architectural periods and styles, though still a residual factor, is nowadays largely meaningless simply because it begs the question of *why and how those 'styles' originated and were applied* in succeeding centuries of architectural development. We are of course talking mainly about:

1 the inherited 'language' of Classical architecture and
2 the indigenous 'language' of vernacular building.

No historical style is infinitely applicable or extendable: it inevitably has to give place to modern architectural forms, stemming from modern needs, modern building economics and modern technologies. Solid external load-bearing walls, supporting 'spanned' flooring, associated Classical glazing patterns and Classical ornamentation, under Classically-pitched roofs with cornices or parapets, can now only be used fully and properly in the context of historical restoration or conservation projects, where no alternative will suit. In modified or restrained fashion, they can otherwise be used (as in the eventual design for the National Gallery extension in London) as an intelligent modern adaptation of the Classical vocabulary for an original building. In some 'populist' commercial office buildings today, ultra-simplified Classical forms or motifs can sometimes be used with taste and success. But, all in all, the legitimate use of Classical architectural forms is severely limited, and not justifiable where other more modern forms (based on modern building methods and requirements) are applicable, or where the 'indigenous' language of truly vernacular building forms and materials is more appropriate.

Of the *preservationists*, whether amateur or professional, we may justifiably say that their legitimate business is directly with preservation, and that they should not seek to impose historical forms or standards upon modern development, except to ensure that new buildings in historic settings must respect in their own way their context, especially where the case cannot be made for slavish adherence to past styles, or the risk run will be that of inferior 'pastiche'.

We come now to the second set of factors bearing upon 'The Great Divide' – again, 'human' perceptions, directly relevant to attitudes on both sides of the 'old versus new' scenario that we recognise only too well. These are, as mentioned above, *the opposing stances which emanate from people's educational, socio-economic class and political differences of outlook*. It would take a lengthy and detailed study to analyse all of these (even if a sufficiently representative sample could be successfully assembled), but it is probably cogent to say that the most telling view of the 'old v. new' problem *ought* to be an educationally based one, in mitigation of the generally traditional British leaning in favour of the 'traditional' in most things. This predilection is reflected from top to bottom across most social strata in this country (in so far as there is awareness of such matters), but from observation it is probably strongest among the upper middle classes, especially those of traditional, public-school background – i.e. the backbone of the Establishment in Britain – where there is not only a strong level of adherence, but virtually a direct identification with the forces of traditionalism, conservation and resistance to change in both architectural and planning matters. Such classes, and therefore such attitudes, not only die hard but continue in varying degrees to be both prevalent and influential.

Yet even the most imperceptive observer of the urban scene could not fail to acknowledge *the pace of economic and technological change* pervading both the governance of life and the new forms of architecture and planning today. Change is now happening at such a rate as would have astonished the most far-seeing of commentators 40 years ago, and which overtakes the sceptics almost daily. The real question is whether this change is properly understood and provided for, or whether it is allowed to happen in an ad hoc or uncoordinated way. We shall be looking below at an example of a modern European state whose planning and development system provides in notable ways a preferable basis to our own.

The Case for Innovation and Progress

The single most pervasive factor in this dichotomy is without doubt that of cost. Both land and building costs have escalated to such an extent in our towns and cities that they increasingly require the use of new technology to meet the various modern demands of speed of construction/development, external walling and roofing, and provision of the necessary internal standards of volume, floorspace, clear working space, lighting, heating, communications, movement, plus access and parking. The cost (and scarcity) of traditional craftsmanship, however, in terms of high-quality architectural 'embellishment' now means that this can only be applied on a selective basis. In many cases, a compromise must be found architecturally between the demands of urban context, traditionalism and financial/technological/developmental considerations.

The architect and developer must be equal to all these demands, and the local authority and the public must also be aware of, and sympathetic to, these problems. In the end, if a compromise is not the right answer, it will nearly always be better to go for a really well-designed contemporary solution, honestly expressing function and purpose, deferring to context only in terms of massing and layout, scale and proportion (and where possible, materials), instead of imposing a pastiche Classical 'fakery' or a crude insensitive commercial 'box', out of scale and just as often bearing no relation to its urban context. This is where the Planning Authority in particular must exercise farsightedness, discretion and sensitivity in dealing with design proposals – and where developers and the public must also take an equally enlightened view. (At the more basic level, the situation is clouded by issues such as our love of amateurism, and our short-term approach to investment in design quality.)

The particular problem which modern 'pluralist' architecture faces, however, is that its more uncompromisingly 'Modernist' design forms are increasingly seen in Britain as acceptable only on island sites sufficiently detached from any historic context, or as part of a separate new development area in which the theme of modern/pluralist architecture has already been set, or alternatively where they can fit easily into an existing scale. Thus we find that the best essays in modern building in this country are often to be found where they are relatively 'free' of restraints of context. Explicitly modern schemes such as Norman Foster's Stanstead Airport and Richard Hordern's Queen's Stand at Epsom Racecourse are cases in point. Terry Farrell's Charing Cross Station, Michael Hopkins' Glyndebourne Opera House and Lord's Pavilion, and Richard Rogers' Lloyds Building are, on the other hand, examples of modern buildings or structures where decisions have had to be made (with mixed success) as to how far to defer to context, or how far to set a new or fresh theme.

Into Europe and the 21st Century

Decisions on design of building schemes of any importance (and in most urban contexts, all buildings are important) normally require *an urban planning policy framework* against which to consider them; they cannot be taken satisfactorily in a vacuum. In Europe today this is of growing relevance, as all countries, West and East alike, experience new or increased building activity as their economies grow, to different extents. Local authorities inevitably have a part to play, in terms of providing either some degree of planning policy framework or design guidance/control, at least in so far as relationships with urban context/planning constraints are concerned. In Britain, a flexible type of system for design guidance (in accord with the national character) has evolved, i.e. where there is a broadly established set of development plan policies from which semi-specific briefs for planning and development are derived, down to specific localised guidance on matters such as shopfronts and signs.

In Germany, by contrast (and again in accord with national character), a much more legally binding and strict hierarchy of planning framework operates, from regional plans downwards, which leaves the developer/applicant in no doubt as to what is, or is not, permissible. As a result, German planners tend to be more specialised and technique-orientated, within a very definitive and tight policy framework, which gives clear and specific guidance in nearly all situations. The outcome of this is that some of the best examples of urban planning in Europe are to be found in Germany,[98] e.g. pedestrian schemes (including public art), high-density housing developments, and building refurbishments. This is also related to the strong architectural background of most German planners and is no doubt also due to a ready access to funding.

The system in Germany (which is based largely on post-World War II national legislation) is that where a detailed plan exists, the landowner can assume that any development falling within the specified use will be permitted (thus, most planning applications are granted – i.e. a building permit is issued). In Germany, an application within a specified land use cannot be refused, whilst one which is not in accordance cannot be granted. Most applications are preceded by negotiations, therefore most are granted. Even where there is no legally binding *Bebauungsplan* (i.e. detailed site building/zoning plan) within the *Flachennutzungsplan* (the preparatory land use plan for the whole municipality, containing both broad land-use principles and site-specific guidance), development can still be permitted, provided it is in conformity with the character of the surrounding area, and avoids conflict with any public interests.

It is also worth noting that, whereas in Britain there are separate town planning and building control systems, requiring separate consents under differing legislation (and which may be both sought and granted separately), it is not so in Germany – one building permit does the whole job. As to public participation, while there is provision for this in the German plan-making process, it is extremely limited in terms of control of development: the general public do not appear to be encouraged, or have the ability, to be involved on a day-to-day basis.

A further feature of the German system is that, in contrast to the rest of Europe, each municipality has the power to embargo all development in a given area while an overall plan is being prepared for that area, which normally takes 1–2 years, with compensation payable after 4 years. During the 'freeze', however, a building permit can be granted subject to qualifications, similar to the Dutch Article 19 procedure. Nevertheless, the embargo provision is felt to have led to increased use of non-statutory plans.

In general, however, it can be seen that the German type of system[99] makes for certainty, confidence, speed and thereby reduced cost, plus accordingly high quality of development. The German system stands largely in contrast therefore to the more flexible British system of broadly expressed, development plan policies generating semi-specific planning/development briefs *plus* the whole British panoply of appeals, public inquiries and public participation, with attendant uncertainty, delays, rising costs and reduction of quality – but preserving the 'democratic right' of developers and public alike to challenge the system at almost every stage and level. It is a luxury which, as part of modern Europe, it is increasingly doubtful that we can really afford. Perhaps the best thing that we can say about the British planning system is that it is infinitely responsive to change!

It is accordingly worth reflecting on the differences between the British and German/European planning systems, as affecting people, development and the quality of the urban environment, down to the actual design of buildings. Whilst some of the more fundamental differences of approach are a result of societal and cultural disparities, and as such are unlikely to disappear altogether, each side of the Channel (still the 'great divide' between Britain and Europe) has something to learn from the other, and the 'ideological gap' necessarily is narrowing. Most European countries, from their regional planning downwards are now attempting to change their planning systems to be more adaptable. In Britain, however, as we advance reluctantly into Europe at the close of the 20th century, we should not be complacent that ours is the best (or only) system. It is evident that we have much to learn from our European counterparts, not least when it comes to questions of how urban quality comes about.

INDICATORS, GENERATORS AND CRITERIA FOR URBAN QUALITY

Assessment of Situation: A General Approach/Methodology

Before attempting to rectify, improve or translate an unsatisfactory urban situation into one that will serve people well and give real and lasting pleasure for the future, we must first identify where not only the weaknesses and deficiencies but also any inherent strengths and advantages lie. In other words, we must draw up some kind of constructive analysis of the situation, in the form of an environmental 'balance sheet' of factors that work both for and against the creation of urban quality, with an eye to what degree of benefit we can reasonably hope to derive from the physical and financial resources available. (We must also take account of the penalties involved in doing nothing, or too little, as well.) Our situation assessment will basically thus be an identification of what actions are needed to achieve or safeguard urban quality, via a perception of outward signs (indicators) and underlying factors (generators).

Relevant criteria

In order to construct this analysis, we must also decide what criteria are most relevant to the urban situation in question, developing our basic approach to these from relevant indicators and generators. There will be some primary criteria which can be applied to virtually any situation, such as that of environmental capacity, which we have already examined in some detail. There will also be ancillary criteria that will apply, according to the particular circumstances, such as (say) a significant presence of old building fabric of high/low architectural or historic value, or the presence of a riverside (or other waterfront) whose existence merits greater attention than hitherto.

Relevant 'indicators' (whether negative or positive) may be in virtually any form, but some *negative indicators* might be:-

- excessive impact of vehicular activity/parking requirements upon the urban scene;
- disruption of architectural cohesion/homogeneity;
- 'bad neighbour' land or building uses evident;
- erosion of previously established (and valid) urban layout forms;
- lack of urban greenery and/or appropriate street furniture/lighting;
- underprovision for pedestrian movement and needs.

Positive indicators might include:-

- a strong local architectural character, in terms of scale/materials/'sympathetic mix'/range of uses;
- evidence of controlled change, as against sporadic and uncontrolled growth;
- sensitive maximisation of existing natural features or advantages, such as topography or water elements;
- balance between commercial and community needs/uses;
- a sense of 'historical continuity' through to the present;
- a sense of order/organisation, and of well-being.

Where *'generators'* (i.e. the underlying reasons) can be established for such indicators, valuable extra clues may be obtained for identifying necessary actions. Identifying the generators themselves will involve local knowledge, research (including talking to local people with direct or indirect interests in the overall situation and any development proposals), and in particular powers of analysis which go further than mere guesswork.

In terms of primary criteria, it is perhaps important to distinguish between those of principally economic, social, physical and 'visual' character (though in practice these may well overlap and interact), since they will broadly comprise the mainly different types. Purists will say, for instance, that 'visual' criteria, such as the townscape type of traditional urban design approach, are too simplistic or even superficial, and do not get to the root problems (such as socio-economic factors or deficiencies) which essentially underlie urban quality questions. Our response, however, is that *it is the interaction of all these criteria that is most relevant for the 'environmental balance sheet'*. Although it is therefore not fully meaningful to take any of these completely in isolation, it must be said that the townscape approach still has a real validity as affecting via the urban character the everyday lives and perceptions of people generally – as is evident from many a public representation (formally or informally) on questions of urban quality. The power, value and relevance of the physical and visual approach in urban design and the environment generally should not be underrated either by academic critics or by 'pure' planning professionals, as long as it is seen as a part, rather than the whole, of the approach to urban situations and problems.[100]

Design Guidance and Civic Design

Following an assessment of environmental needs and qualities, as above, the next stages will usually involve (to some extent or other) one of the following:

- creative physical planning (as in urban or civic design schemes)
- design guidance or control.

We discuss these in very broad outline below.

Creative physical planning

Urban design schemes, in the form of 'civic design' projects (after a period of public disenchantment following those very 1950s and 1960s schemes of box-like, geometric blocks set in windy open spaces, adorned by overhead walkways of dubious merit and appeal) have now re-emerged in rather more intelligent and sensitive urban layouts. Now the accent is on 'humanisation' via scale, landscaping/greening, and relationship with existing urban character, but also with imaginative use of 'theme', in ways that contribute positively and substantively. Urban designers are finding an increased need for their skills in 'civic layout', not only in Britain, but particularly in all the more affluent countries of the Western world, where the application of urban design techniques and principles can yield rewarding results.

In Britain, the use of urban/civic design skills has been found to be especially applicable in large cities such as Birmingham or Manchester, where change on a larger scale than normal is called for, due either to new economic opportunity or to the urgent need to renovate areas of civic importance that have been consistently neglected. Opportunities of this kind also exist in, and are relevant to, towns of more modest scale but of particular function or interest, such as seaside towns like Margate in Kent or Brighton in Sussex, where the use of urban design skills in seafront/promenade situations can have a cohesive, and therefore beneficial, effect on the town's fortunes.

Design guidance/control

Design guidance or control is a different matter altogether, raising completely different issues and implications. As we have discussed earlier in our examination of planning and development in Britain, there seems to be nothing which raises the attention and expectations of the public, and often the resentment of private sector architects in this country, so much as *the vexed question of design control by local authorities*. Whilst public debate of broad overall urban or civic design layouts may be accepted as part of the 'democratic process' in Britain, the criticism or amendment by planning committees or

Planning Officers of design proposals for individual buildings can arouse a quite disproportionate degree of antipathy on the part of their designers. The developer will not be slow himself to accuse the planners of 'tinkering' with the scheme (especially if the delays to getting the development 'off the ground' incur added costs).

The answer to this latter problem, as we have said previously, would appear to lie in pre-application negotiations based on the production and use of design guidance by the local council, from leaflets on specialised subjects (e.g. shopfronts and signs) to individual planning briefs or comprehensive design guides, such as those produced (notably) by Essex County Council in the 1970s, or Surrey County Council in the 1980s for residential development. (Hove, in East Sussex, is a good example of a district council which published in the 1990s a whole range of design-oriented planning leaflets of a high order of quality and content).

By contrast, however, North American (and German) public sector planning practice does not allow detailed types of design problem to arise. Reliance is placed instead upon the use as necessary of *planning briefs,* which incorporate outline guidance on design in a most general way, related to the use of land, desirable densities, broad architectural massing/scale, and building materials (but only so far as appropriate), whilst allowing designers the maximum permissible flexibility as to actual detailed form of the building(s) proposed. If the designer proposes something that infringes the light or access to an adjoining building, that will then be a matter which lies more within the remit of legal redress than of planning authority intervention or control. In the USA, this form of control principally via planning briefs goes either under the name of *design review,* or *site plan review.*

Site plan review in America can be instituted as a mandatory development requirement for the entire community, or the downtown area, or only in special instances. For example, it can be activated where increases in height or density are requested over what is permitted by the existing zoning or where a conditional land use is proposed. In either case, standards and criteria for approval must be carefully documented in advance. These standards must also be applied both consistently and objectively.[101] A notable example of an American development carried out via (unusually) a *'design code'* type of system is 'Seaside' in Florida, where the code specifies only the proximity, materials and colour range for the houses (holiday and retirement homes) to be built there – plus one or two compulsory requirements such as the use of porches (of whatever form) and white-painted picket fences (not plastic, which is discouraged generally). Other than these few restrictions, the code is very much in favour of individual imaginative design features for each house (such as towers or gazebos), though the successful use of these design features would in turn tend to reinforce the case for using architects and skilled, sensitive local builders. The density and proximity parameters laid down for the site ensure a pedestrian 'walking' scale for the development, whose concept has been likened to that of a Classical or hellenistic town, set down on the Gulf of Mexico. It may perhaps be argued that Seaside is also a prime example of an elitist development for wealthy business people or retirees – but if money used intelligently produces a high level of urban quality, we can scarcely quarrel with it.

On the other hand, if the link between ample resource funding and urban quality were to be seen as exclusive of all other means of producing the desired result, we would then point to the wider need for requisite planning and design powers to be widely available and applied, in the interests not of a single enlightened or favoured sector, but of the whole community at large. This is perhaps an advantage of the British planning system, which enables a range of planning and design advice and/or

Figure 37 Cedar Rapids, Iowa: enhancement of a US regional city.
The single worst problem in Cedar Rapids is the lack of a traffic-free environment. Public/private partnership, key project implementation, and marketing publicity will assist in the realisation of potential for vibrancy in this regional city.

development briefs relating to its development plan provisions to be produced as necessary (in a way not only geared to the operation of the private commercial sector but also to the perceived general interests of the community).

Public and Private Sector Co-operation

In short, the aim should be to generate development that is 'democratically beneficial' as

PERSPECTIVE ON URBAN QUALITY

An urban waterfront promenade is proposed; a sophisticated pedestrian activity corridor can be created in what is now a service/parking zone. The promenade would be predominantly paving, with canopy trees in grates and provisions for flexible seating.
Source: LDR International, Inc.

opposed to 'individually assertive'. In theory, this should be the outcome of the current situation in Britain, where there is increasing reliance happily upon co-operation between the public and private sectors as to development generally. This co-operation may come in any one of a number of forms, including:

1 Joint development participation by the developer and the local authority, arising from jointly agreed aims and joint funding arrangements, without government grant aid.

2 As above, but with government aid, via programmes such as City Challenge or Rural

208 PLANNING FOR URBAN QUALITY

5-minute radius

10-minute walking radius

Lawrence Tech

WXYZ-TV

Duns Scotus

Southfield City Centre

PERSPECTIVE ON URBAN QUALITY

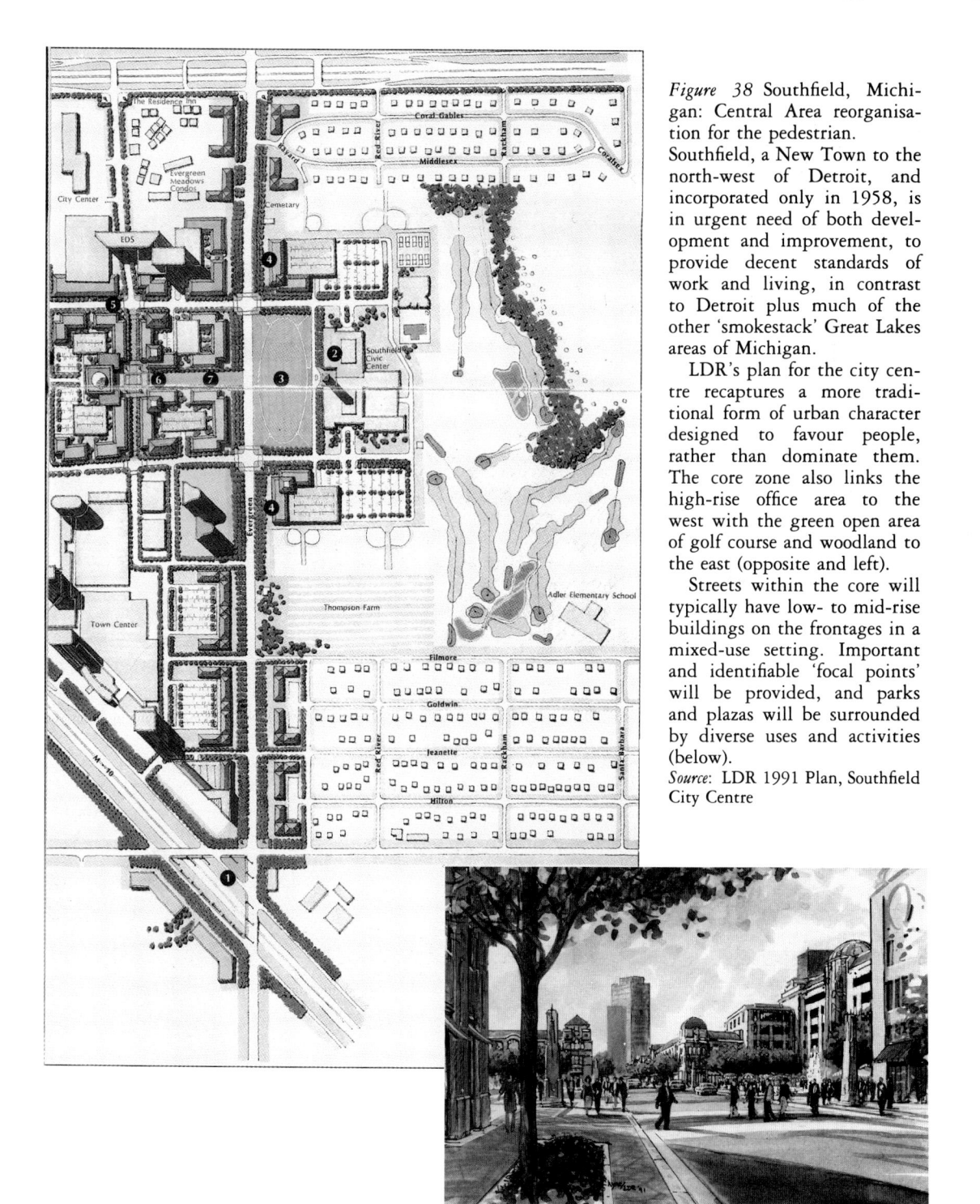

Figure 38 Southfield, Michigan: Central Area reorganisation for the pedestrian.

Southfield, a New Town to the north-west of Detroit, and incorporated only in 1958, is in urgent need of both development and improvement, to provide decent standards of work and living, in contrast to Detroit plus much of the other 'smokestack' Great Lakes areas of Michigan.

LDR's plan for the city centre recaptures a more traditional form of urban character designed to favour people, rather than dominate them. The core zone also links the high-rise office area to the west with the green open area of golf course and woodland to the east (opposite and left).

Streets within the core will typically have low- to mid-rise buildings on the frontages in a mixed-use setting. Important and identifiable 'focal points' will be provided, and parks and plazas will be surrounded by diverse uses and activities (below).

Source: LDR 1991 Plan, Southfield City Centre

Challenge, or alternatively via the Single Regeneration Budget, in which 20 programmes from various government departments have been drawn together, enabling funds to be made available locally through a single government office, and focused according to identified local needs and priorities without being 'earmarked' from the centre and thus inflexibly administered.

3 Joint action between either a private developer, or in some cases a local authority, and English Partnerships, the urban regeneration agency newly integrated in the early 1990s to promote high standards both in its own developments and in those that it supports. The Agency ensures that all its projects support long-term environmental enhancement, via good design, sensitive landscaping, tree planting and suitable aftercare. It also has a special role in promoting the reuse of derelict land, and the reclamation of contaminated land. Similarly, the Rural Development Commission, another central government agency, backs local efforts to promote the economic and social well-being of rural communities.

4 In terms of housing, whilst the accent during the 1980s in particular (and into the 1990s) has been on housebuilding for private ownership (with 67% of houses in the UK now owner-occupied), efforts have been made in recent years by landlords and developers to improve the standard of private rented accommodation. Attempts are also being made by local councils and housing associations to provide a decent amount and standard of public sector housing, either as rented or 'affordably priced' housing for sale (dependent on availability of suitable

PERSPECTIVE ON URBAN QUALITY 211

Figure 39 Lakeland, Florida: lakeside potential maximised for community.

LDR's Development Initiative for downtown Lakeland in Florida, between Orlando and Tampa, maximises the potential of its lakeside setting for community benefit/amenity generally, including outdoor concerts and gatherings, with a waterfront promenade and park, hotel, restaurants, and other commercial developments and plazas. The Americans have not been slow to exploit spectacular settings such as this, and their determination and speed in implementing such schemes are enviable in many respects (opposite and below).
Source: LDR 1990 Development Initiative, Lakeland

land or sites) to those unable to pay the overambitious prices which still tend to be the hallmark of the private housing market. Community-based schemes such as Estate Action, Housing Action Trusts, City Challenge, and Safer Cities, also help to refurbish rundown housing estates and to regenerate neighbourhoods into desirable living areas.

5 A further form of 'joint' action may be said to comprise the situation where the local authority may derive 'planning gain', in the shape of local environmental improvements, as a 'spin-off' from a commercial developer's scheme – provided always that the 'planning gain' element is physically relevant to the development site and the content of the scheme to be implemented on it.

Local/central government administrators and politicians now recognise the importance of a high-quality environment in attracting inward investment generally, and perceive that this investment helps improve the environment further, as described in the Department of the Environment's 1994 discussion document *Quality in Town and Country*. The publication of this document received widespread acclaim amongst professional planning institutions, associations, and others. It has left open, however, the question of finding ways of 'drawing that investment into the areas where the benefits are greatest and the improvements most needed'. This does not of course invalidate the useful contribution it has made in clarifying the agencies for obtaining urban quality in an age heavily impregnated with the view that market forces affect everything and will therefore solve everything – a patently false (or at least inadequate) premise, as the application of an in-depth environmental analysis, such as we have outlined above, will readily show. The 1980s scenario of the 'marketplace supreme' now gives way to the more informed view of the late 1990s.

SUSTAINABILITY IN URBAN QUALITY

As we progress into the 1990s and beyond, the one main concept which has emerged at the start of the decade bids fair to become the single most prominent environmental and planning issue for our time – that of *sustainability*. In order to understand this term in all its connotations, we will look at a wide span of history of urban building forms and techniques, with their social, fiscal and economic frameworks. First, however, we must see how the modern-day concept of 'sustainability' has arisen, and how the term is generally defined, via the provisions of the mechanism know widely as *Local Agenda 21*.

Local Agenda 21

What is Local Agenda 21 (LA21)?

In 1992, the nations of the world met in Rio de Janeiro to discuss the state of the world's environment. This was known as the Earth Summit, but sometimes it is simply referred to as the 'Rio Conference', or just 'Rio'.

As a result of the summit a document called Agenda 21 was produced. It was a list of measures intended to address problems like global warming and infant mortality, and the crisis over food, fuel, water, poverty, disease, livelihoods, etc.: in other words 'Quality of Life'. It has been defined as 'a comprehensive blueprint for the global actions needed to effect the transition to sustainable development'.

Agenda 21 (and therefore Local Agenda 21) has three areas of concern:

- increasing and protecting social equity, and equality of opportunity;
- safeguarding and improving economic security; and
- protecting and enhancing the natural environment.

It was recognised that in order to achieve a better quality of life, changes would need to be made at the local level by individuals, groups and local government, hence Local Agenda 21 (LA 21). Further, that this can be achieved by the empowerment of ordinary people.

Local authorities accordingly have been called upon to consult and involve local people in order to draw up a local plan of action. The aim is to achieve 'sustainable' communities.

What is 'sustainable'?

'Sustainable' means:

- 'meeting present needs, without harming the chances of the next generations being able to meet theirs';[102] or
- 'not cheating on our children';[103] or
- 'living off the interest, not the capital'.

It is thus felt that commitment to the principles of LA21 will lead to the establishment of sustainable communities. This will restore vitality to our villages, towns and cities by encouraging people to work towards common interests. It will not happen quickly – it will take decades; but a start has to be made because our resources are disappearing and the world's communities are rapidly falling apart. We look therefore at the roots, then the mechanisms for sustainability as far as urban quality is concerned, with reference to the UK and elsewhere.

URBAN QUALITY: OLD AND NEW NEEDS

The Roots of Sustainability and their Relevance Today for Urban Quality

Until comparatively recently (probably the last half of the present century) most building effort was directed to 'permanency' in the building fabric. Even low-cost artisans' housing was built of durable brick, stone, slate, tile, thatch, render and seasoned timber from virgin northern forests. Heating by means of open fires and simple convectional systems, together with traditional lath and plaster insulation, and small glazed areas satisfied simple social needs. More important buildings, often located in town and city cores, were traditionally constructed in time-honoured systems to last for many hundreds of years, or even 'for ever' in the minds of skilled labour forces, using proven techniques of construction, quite often from Roman and medieval times. Labour was cheap, and until the 18th century much building finance was obtained from 'old money' which saw ultimate value in durability, rather than a quick money profit based on limited liability and aesthetic responsibility.

In Britain, and most parts of developing Western societies, public health reforms in the late 19th century led to minimum standards of building construction and design being required by law, and these minima rapidly became maxima as competition to make money grew explosively, like urban populations, from that time onwards. The end of Victorian prosperity, and the effect of two world wars, which produced far-reaching new technologies and a curtailment of that old prosperity, carried the additional problems of social revolution requiring higher living standards, in a still burgeoning population. More energy to heat, light and ventilate buildings grown smaller (which were quicker to finance and construct) and more mobility by way of the oil engine, all drove towards the present worldwide ethic of a 'disposable environment' and its inevitable corollary of an architecture that itself is disposed of, as its material needs extract the fossil fuels and mineral assets of its base.

We have now arrived at a critical point in the perceived linear progress of building technology and planning standards where 'cosmetic solutions' have become patently unsustainable. Political 'short-termism' is hardly useful in its resolution, since the latter's application lies too closely with shorter-term financial consideration, in much of the world today. This should neither surprise nor alarm us into running backwards to concepts of the urban or indeed general built environment of improbable pasts; nor should we grasp at impossible futures, for an urban quality which certainly will not be 'resolved' in this way. We appear to have been panicked by the speed of social and technical changes forced on us (by ourselves) since the end of the 19th century, and we must learn to 'change with the changing' in an intelligent manner.

Clear, firm and subtly simple agents of *sustainable improvement in urban qualities* are available to us at the present time. Several of these were outlined in Sir Richard Rogers' Reith Lectures broadcast by the BBC in February/March 1995. The comprehensive re-introduction of safe, convenient, clean, fast public transport across city centres at affordable cost, is already growing quickly across Europe.[104] The significant reduction of oil-drive private cars through the urban environment is urgent, and the technology to achieve this is with us. Computerised systems of traffic observation and control now exist, and the application and maintenance of significant 'green' areas of real use to populations keen to 're-urbanise' for social, business and cultural reasons are being installed as an accepted part of the new city and town, across the developed and developing world, where leisure time is growing.

The fiscal framework to enable such changes has yet to be put in place, both through the design of the buildings, the energy they presently use, and the installation of all the road and transport management requirements in contemporary city use. A departure from old accounting systems, familiar to the Roman world of agriculture, is long overdue, and the true value of 'cost in use' and long-term sustainability, must replace short-term return on invested capital when the future life of built urban fabric is planned for the generations to come.[105] The Victorians, in their own ways, did as much for our own present, and we should at least repay the debt.[106]

Sustainable Revisions in Land Acquisition and Construction Realisation

All land is finite, and that at City centres has been driven to the highest financial values, as world populations have grown and concentrated in response to social, commercial and historical pressures. In most Western legal systems, buildings and other constructions in, on, over or under land, have been traditionally regarded merely as 'fixtures' to their land sites. The intrinsic value of much urban land with redevelopment value can thus be seen to be greater than sites occupied by depreciating structures of conventional form or use, as they continue to be in demand for commercial, industrial and population activities. (One of the authors, as an architect, has a direct experience of building four times, from foundations upwards, in one professional lifetime, on a single urban site). Such redevelopment might well today involve open space for leisure and health use, or even new public transport needs. In most cases, the purchase price of property will reflect the value of the land potential of the site, rather than the 'replacement costs' of an obsolete structure of little worth in its current use, or in its 'heritage value'.

This approach will necessitate a much deeper examination of property values than 'current market value', and 'planning gain' will have to have a real meaning for owners of land over-ripe for redevelopment in modern terms in urban centres. *A participation in the future positive value of a redevelopment site* ought to be made available to freehold sellers in a similar manner to pension funds, at least, and this could well attract small owners to dispose of strategic small packets of land in an efficient

Figure 40 City centre public transport, Strasbourg.
The European city of Strasbourg, France, exhibits a highly sophisticated approach to its central area public transport provision. Not only does it provide 'trams' of ultra-modern design for its citizens and visitors, but it has also built a central terminus feature to match this adventurous level of design and technology. It thereby mirrors the resurgent European spirit of looking forward to the 21st century, rather than back to the past.
Source: Michael Parfect

and fair manner, at affordable cost to development agencies. In turn, sound local planning policy, duly recognising the importance of quality in rebuilding old urban environments, might well be strengthened by its requirement for the installation of buildings of true architectural worth and sustainability at the city core.

Most professional architects, have, for some time in Britain, been acutely aware that buildings consume about half the energy (and produce about two-thirds of the waste) on a national scale; that they also suffer rapid change of design use as society and technologies change with increasing speed; and that under present contractual procedures, they are relatively expensive to realise and finance, especially where high quality is envisaged. *The whole economic and craft base of building technique has changed, and the structure of land ownership has become both commercialised and state regulated*, to a degree unimagined by our forebears. Social order, or, perhaps even disorder, in many of our city agglomerations, demands an urban quality that owes very little to past environmental and technical experience. The motor vehicle, the economics of food and goods distribution, noise and atmospheric pollution, computerised information transference worldwide in 'real time', together with an almost negligent general acceptance of the liberty of the individual, at the expense of social order, are all factors to be urgently addressed, if urban life is still to be civilised in nature. There is little doubt that humankind has a remarkable ability to adapt quickly to change, and modern techniques of communication, information transfer, exchange of ideas and instruction, facilitate this ability once a problem has been recognised, or once an advantageous opportunity has been realised. But change must be beneficial and acceptable generally. There is now hope that scientific technique, once properly ordered by moral and ethical force, and intelligently applied, can win dramatic improvement to urban quality in sustainable terms.

Lifetime durabilities could be a worthy aim to pursue, in the search for an acceptable and sustained decent standard of world living. The fiscal system, which controls the provision of capital to virtually the whole of our production and consumption systems today, is operated by relatively few institutions under the control of few people, and it is becoming evident that the resources and skills required to run democratic world capital systems are failing. Outdated accounting methods often produce bizarre crises in financial dealings at an international level, while 'waste' of investment capital, in the pursuit of quick returns, is leading ever more frequently to uneven supplies of money and the uncontrolled exploitation of whatever medium lies conveniently to hand, for providing short-term relief. Rain forest timber, ocean fish stocks, overdependence on fossil fuels, meagre investment in waste processing, are some of the obvious signs of incompetence in world financial institutions, and the *construction industries, which are high consumers of initial capital, are often compelled to carry one of the heaviest financial penalties for this short-termism*. This has probably been true, since the time when bricks were made without straw in Egypt, or the Parthenon builders economised by resorting to hollowed marble facings, but the Gothic cathedrals gained the fourth dimension of time, over their long extended construction periods.

Status and Deficiencies of the UK Construction Industry

There are practical points to be argued both for and against the performance and contribution of the British construction industry to architectural and building standards in the UK today. One might claim on the one hand, as above, that the industry has been the victim of short-termism arising from a fluctuating degree of financial confidence in government and its economic management/fiscal policy, affecting in turn the pattern of consumerism for both house building and commercial development.

An additional factor clouding the issues is the conflict over public acceptability of traditional architectural 'styles' as against modern technological advances in building construction and forms. On the other hand, one may point to the chronic design backwardness of the British housebuilding industry, which lags behind European and American practice in certain respects (e.g. the British builders' practice of perpetuating the pseudo half-timbered house or Neoclassical ethic on 'customer demand' grounds, but failing to plan functional elements such as garages and workrooms below ground floor level in private houses as in the rest of Europe, in a era when development densities and housing land supply in Britain are very much acute issues).

Equally, the low productivity and technical standards of the housebuilding industry in Britain co-incident with the migration of skills to machine-tooled industries may be said *to stem also from builders' access to high land profits*,[107] which has meant they have not been under much competitive pressure to modernise. This is reflected in the high purchase prices of new houses and the practice of 'land-banking' by house building firms (creating an artificial extra pressure on the 5-year supply of land with planning consent or otherwise identified for housing development, required from local councils irrespective of environmental capacity constraints, via the development plan system).

The development plan system in Britain by and large caters adequately for the housing industry in terms of land identification or 'supply' (despite the unrealistic demands imposed upon it by central government) but for all that there are recurrent complaints by house purchasers of lack of supply of affordable/low-cost housing at the same time as complaints by the construction industry of inadequate operating margins.[108] This leads back to the point discussed earlier, as to the importance of quality and value of buildings, in relation to value of land, in which the former will comprise an increasing proportion generally.

The building industry in Britain merits close consideration, since the quality of building *per se* is as important as the quality of design in the urban environment, and as we have said has connotations of sustainability. In an earlier section we noted that many of the high-rise, systems-built failures of the 1960s–1970s (in particular, the disastrous Ronan Point collapse) which stemmed from the pressing need to house large numbers of people on relatively small areas of land, were in fact largely due to the inability of the British construction industry to build to the required standards, for which, however, the local authority architects were obliged to take the brunt of the blame, on behalf of their political masters.

The problems of the British construction industry go hand in hand with the predicament of the British architectural profession. Both are highly vulnerable to national economic fluctuations, affecting both standard of product and workforce employment. The migration of skills elsewhere, in the face of the time/money imperatives of both design and building pressures, have accentuated the growing recourse to modern 'industrialised' forms of building, designed increasingly via CAD (computer-assisted design) techniques, with major elements fabricated in redundant shipbuilding yards, thus increasingly replacing 'traditional' architectural styles and construction. Whether or not they will prove 'sustainable' remains to be seen. New forms of building design will continue to be rationalised in this way, and where environmental context and other planning considerations permit, we may expect to see greater numbers and diversity of these new forms in our midst.[109]

The Husbandry of Land

We live in a time when the strictly limited and finite resource of land threatens to run out all too rapidly in Britain unless both planners and central government face up to the question of land recycling on a major scale, to a pattern that allows urban restructuring and reuse. As the President of the RTPI, Michael Welbank, said at the Institute's annual conference in June 1992, 'we might even reach the position of determining the absolute limit of the amount of developed land in this country, and when that's reached we rely on recycling land thereafter'.

He warned politicians not to expect the planning profession to solve the sustainable development dilemma as if it were a minor technical matter:

> It is a profound problem in which we wish to play our part, and we should all be addressing its solution far more vigorously than is currently the case. . . . *we have the dangerous situation where the environmental demands of society are running ahead of our capacity to deliver.*

He complained that 'planners are being asked to fight this battle without weapons' and that consequently they have no accepted coherent or even theoretical basis for action.

At the same conference, however, the *sustainability theme* was also taken up by the Planning Minister, Sir George Young, who saw the achievement of sustainable development as a major challenge to the planning system over the next decade. He stressed that *the plan-led system had a key part to play in securing environmental improvement*, 'for if plans take environmental considerations fully into account, decisions made in accordance with those plans will in themselves be green'.

The need to ensure that local planning authorities were meeting their sustainable development obligations (underlined by the government's Planning Policy Guidance Notes PPG1 and 12) was taken up also by the Council for the Protection of Rural England. The CPRE Senior Planner Tony Burton pointed out that development plans will now have to include policies on environmental improvement, transport management and the like, while *environmental capacity concepts* are likely to edge out household projection techniques as a basis for accommodating housing development. 'Planners must no longer seek simply to "balance" the needs of development and the environment, trading one against the other', he warned.

He emphasised that in a sustainable world, the urban or rural environment will come first and act as the guiding framework for all development. We should plan to live within our environmental means and reduce our demands on scarce resources. He said:

> You must grasp the nettle of demand management. The days of predicting the demand for limited resources, and then using the planning system to ensure their provision are ended. You will need to examine development proposals to see if they are efficient in their use of energy, water, land and minerals.[110]

Notwithstanding the fact that all these comments were being voiced concurrently at the United Nations 1992 Conference in Rio de Janeiro on the protection of the Earth's resources, and improvement of the environment, *it is clear that planners will have to tread carefully in framing policies so as to gain wide acceptance for them within the democratic process.*

MECHANISMS FOR URBAN QUALITY IN A SUSTAINABLE FUTURE

At present, local authorities are finding it difficult enough to perform their basic statutory roles without taking on much else in the form of urban design studies frameworks, or guidelines. Furthermore, most planners by virtue of their training and organisation tend to see local plan-making in two dimensions rather than three, and often only in terms of land use allocations, mixes and quantities. What is needed to assist Planning Authorities is some kind of computer-assisted process[111] which enables plans to be translated into three dimensions in key areas including details and colour representation. Increasingly, the initiative or spur for urban design, as a major means for obtaining urban quality, is coming from local alliances of people who feel that if they do not create or press for a vision for their area, no one will. One small but important body, unique of its type, which has taken up the challenge of a dire lack of official support in Britain for the promotion of urban quality, is the Urban Design Group.

The Urban Design Group and its Aims

The impetus behind the creation of the UDG in 1978 came originally from a gathering of architects and planners who were concerned about the meeting-point of their two professions. Its members now come from a wide range of backgrounds, but that concern remains. Though a definition of urban design is elusive, most people would agree that its emergence among environmental professionals could only be a good thing. Main planks in the UDG's platform are:-

- A refusal to hand over the environment to the care of any single profession;
- a concern for the public realm of towns and cities; and
- an insistence that remaking the environment should be a popular activity.

The relevance of these aims as a vehicle for sustainability is evident.

Despite the lack of paid staff, and difficulties therefore of administration, the UDG has grown in membership and runs an impressive programme of events, including exhibitions, competitions, workshops, seminars for local authority officers and councillors, an annual conference, the publication of *Urban Design Quarterly* for a widening audience, and providing some CPD (continuous professional development) services.

The UDG Chairman in 1992, Jon Rowland, when asked to define what he would like the UDG to focus on, suggested 'Finding out the best mechanisms for revitalising towns and cities'. For several years, he has held the view that the UDG should take the plunge and become a sort of professional institution. Its lack of an institutional bureaucracy means that it will survive in its original form only as long as the enthusiasm of its members, therefore its relationship and links with other professional and amenity bodies are probably vital.

Fortuitously, in March 1993, the incoming RTPI President and Civic Trust Director, Martin Bradshaw, in a speech to a meeting of the UDG, urged professional planners to get together with the Urban Design Group, and perhaps the Civic Trust, in a liaison designed to put the environment back at the top of the political agenda. He argued that there was a need for 'a mouthpiece for urban areas' – perhaps an institute for the urban environment, a city and towns commission, or even a campaign for urban England. The proposed three-way link-up, he said, would look at ways in which design resources might be made available to local authorities at the plan-making stage. (A first step would be for the three organisations to meet and set an agenda, including an immediate dialogue with the local authority associations). Planning's contribution to urban design requires strengthening, Bradshaw declared. At the national level, the RTPI and the RIBA should join forces to raise the profile of design concern.

Community Development Trusts

In Britain today, we appear to be locked into a tight, vicious circle of emasculated local authorities, market-driven public development companies, militant local political pressure groups, well-intentioned amenity bodies, and a perplexed, irate, and generally uncomprehending public, to many of whom the mere mention of development, design or planning is like a red rag to a bull. *A major underlying problem in all this is that change in our urban environment so often takes place without a real share or contribution on the part of those actually involved in the end product.*

The most common type of 'case for treatment' in Britain nowadays is the area of environmental decay or urban deprivation. This often comprises one or more areas of architectural or historic character which have seen better days, but which somehow may continue to have a recognisable community life, usually where the essential nature of the area is still mainly residential, even if with commercial elements.

Community Development Trusts have provided a model of sustainable, community-based regeneration since the 1970s, but only 20 years later did they begin to enter into government agendas. Most trusts start off by being dependent on public funding but have in common many of the same characteristics, 'combining idealism, community support, and a pragmatic search for viable and sustainable projects to meet local needs' (as a special feature article in the June 1992 journal of the Town and Country Planning Association put it).

The North Kensington Amenity Trust in London is generally thought to have been the first and most successful trust of its kind. It was launched in 1971 with a council grant of £25,000, and turned 23 acres of derelict land under the Westway flyover into a £20 million asset supporting 750 local jobs. The formation of such trusts was very much in the idealistic spirit of the 1960s, but also reflected the realisation from the mid-1970s onwards that local authorities could not or would not continue their previous role of carrying out responsive redevelopment. By being part of the voluntary sector, trusts were also to experience, however, the uncertainties of core funding from the public sector and the difficulties of obtaining the confidence and support of the private sector.

The Hastings Trust in East Sussex has had much the same experience, except that the project there relied almost entirely on public funding. The Trust in fact first manifested itself as the agency operating the Hastings Urban Conservation Project, which was established in 1986 by the East Sussex County and Hastings Borough Councils, to encourage the regeneration of architecturally rich, but financially threadbare Hastings and St Leonards. The two councils put in £425,000 over the 5 years up to 1991. This total was raised to over £0.5 million by grants from English Heritage, with only £45,000 in additional funding from sponsorship and sales. The Hastings Trust, which was set up to build upon the work of the Conservation Project, was obliged to reverse its funding basis almost right away in trying to secure private sector finance, owing to the problems of insecure public sector core funding undermining the impetus of the project in terms of maintaining basic community services. Thus Hastings has suffered the simple irony that 'a town which is run down and thereby could benefit from the work that the Trust could do, cannot subscribe to that work – because it is run down'.

The unknown factor, as ever, is the extent to which local authority funding is underwritten, or frustrated, by central government. There is some evidence that the latter is giving increasing attention to how Community Development Trusts (*inter alia*) might play a bigger role in regeneration partnerships, and thus in the process of obtaining greater urban quality. What is still lacking, however, is any kind of consistent overall strategy for the support of voluntary and community sectors in urban areas. While being generally supportive of the idea of 'partnership' for the achievement of urban regeneration, central government has yet to give clear guidance on how the concept might be developed or applied, though more attention is now being given to the need for financial co-operation between public and private bodies. A regional, rather than central government, basis holds out more prospect for realising this, however.

Despite this, or perhaps because of it, Trusts have made much progress in developing their own role at the local level, even during a period of severe recession, and despite problems of funding sources and legal status. Community Development Trusts are just one part of a larger range of diverse and independent community enterprises which have common aims of self-sufficiency; local control; partnership; and economic, environmental and social renewal of a closely-defined area or community. *The time has come for much greater support and guidance from both central and local government on how this sector can be encouraged to attain its full potential and thus to further the aims of sustainability.*

NOTEBOOK: SPECIAL CONSIDERATIONS

If we are to be true to the present needs and concept of a *'pluralistic society'*, as relevant to sustainability, the common approach must embrace aspects such as 'inclusiveness' (catering for neglected sectors of society, namely women, people with disabilities, and children) and thus 'diversity', where these are perceived to be lacking. We shall therefore briefly consider two of these aspects, namely women in society and their place in planning as a main consideration, together with questions of provision for certain special or minority groups additional to mainstream population in Britain, and other special considerations under the subheading of 'Notebook', though these neglected or underrated social aspects in fact demand more specialised study than we are able to give here.

Provision for Women in the Environment

This has traditionally been a neglected aspect in the planning and design of our urban environments, and only recently has greater attention been given to the special needs of women in society. A landmark paper *Planning for Women* was produced by the RTPI as their Practice Advice Note No. 12 in January 1995, from the original 2-year study. Reference to this work is particularly recommended. PAN12 covers legislative aspects; access to services; design; mobility; and the planning process. We highlight below some factors from it of general relevance to urban quality matters.

- *Generally* the modern working woman/single parent, the family woman and the elderly woman/man alone form three distinct types of urban dweller, all needing greater convenience in their interfaces with the urban function, if quality of urban life via planning is to be improved upon.
- *Convenience shopping (as distinct from comparison shopping)* must be near at hand, preferably within pedestrian reach, for the two latter types above. The out-of-town shopping mall, however well stocked, is not easily accessed by those who do not drive cars, and there is thus an urgent need for the return of the convenience corner shop, for day-to-day needs, provided near at hand. (The government's pledge to halt the spread of shopping malls in February 1995 was welcome, but unfortunately long overdue.)
- The first two types urgently need *child-minding facilities*, if long journeys to larger, distant services are to be accessible.
- *Present multi-storey and some basement public car parking facilities* are barely acceptable to many single female users. They must be better lit, serviced and supervised, to improve popularity, and a revised planning consideration must be given urgently to the interface of public and private car parking facilities, within safe, dry access to shopping and other services.
- *Clean and reliable public transport* must be readily available to women users, especially those without cars.
- Outdoor, and weather-protected, *play/recreational areas for children and adults* must be within easy reach of necessary user facilities, and be *well supervised* (at least as thoroughly as street car parking by traffic wardens). The security of the young and the elderly in large open areas, without supervision of heavy public use, is now compromised in high unemployment times, and the easy casual access from adjoining distributor road systems often existing adjacent to open spaces, is an added hazard.
- *Evening and night-time recreational areas* will only appeal to women in all categories (especially the first, where many will be unaccompanied) if they are clean, well lit, attractive, busy and well supervised. This ideal is often unrealised today in the city which has become dead at night, due to the outward flow of commuters, and the scattering of cafe, restaurant, theatre, concert hall, late shopping and cinema sites.

Further relevant publications

Matrix, *Making Space: Women and the Man-made Environment* (re multiculturalism, and design discrimination against women), London, Pluto (1984); *Changing Places*, London, Greater London Council (1986); *Planning for Choice and Opportunity*, London, Royal Town Planning Institute (1989). *Access for Disabled People*, RTPI Practice Advice Note No. 3 (1988); Publications by the Women's Design Service (via RTPI Library).

Racial, Ethnic and Religious Groups

These groups represent an equally neglected/underrated facet of our society. Since the end of World War II, there has been a significant and varied influx of such groups into most parts of Britain. Because of better employment opportunities and the natural wish of individuals of similar origins to keep in close touch with their social and relationship peers, many British urban environments have significant clusters of African, Caribbean, Asian and new Commonwealth groupings in older, more central, and therefore often lower-cost areas of our cities.

Post-war immigration from various parts of Eastern and Western Europe has been relatively easily absorbed in prosperous times, and along with an established and influential Jewish component (since at least the Middle Ages) it has formed the distinct national character. Immigrations, mostly from Eastern Europe, at the end of the 19th century and also much earlier, from the Low Countries, produced an even smaller impact on the racially and religiously mixed existing populations of England, Scotland and Wales, and on the whole, were assimilated into the British cultural and class system quite easily.

The urban fabric (with the exception of religious buildings in the larger towns and cities) has shown relatively little of these immigrations, and today, this can be seen as a shortcoming, in the quality of the urban scene. The vigour, colour and freshness of much of the ethnic and religious art and architecture now available from immigrant groups is to be welcomed in our cities and towns in prime positions, where quality manifestations will be appropriate. Mosque, temple, synagogue, ethnic centre, pleasure garden, specialist library sincerely realised and maintained should form a focus for celebration in the community at large, and demonstrate the confidence and faith in Britain's multi-racial, multi-cultural and multi-religion new society.

THE WAY AHEAD IN PLANNING AND DEVELOPMENT

What is particularly clear is that the benefits of private sector finance and development expertise are essential for a revitalised public realm environment, in which the public sector/local authorities need also to play a constructive and active role. Necessary in turn to this is not only a relevant and appropriate planning system/context, geared to ensuring that the right things happen in the right places and at the right time, but also an even-handed and constructive approach to 'fiscal energy' programming by central government. Only in this way can the provision of urban quality be stimulated over the environment as a whole. Finally, the whole question of urban quality must be set in a context of the present and future needs of sustainability.

One of the best English language works yet written on the subject of the urban environment in an analytical but thoroughly stimulating and engaging way, is Michael Middleton's 1987 book *Man Made the Town*. This fascinating and hugely diverse work positively abounds in cogent examples and in perceptive thought about the circumstances, forces and influences which must be understood and harnessed if we

PERSPECTIVE ON URBAN QUALITY   223

BRISTOL

Figure 41 In conclusion: towns and people: Bristol/Sarlat.

The view from the high, narrow street to a panorama of 'the city in the valley' has been an excitement since cities began; the experience of passing from one world into another, through the narrow, darkened opening, to a sunlit mystery beyond; the thrill of moving smoothly along a curving path to a surprise yet out of sight; the liberation of the mind, when the view from a confined environment gives out to landscape, townscape, or waterscape; the secure feeling of architecture flanking the confines of a shaded (or brightly night-lit) urban street filled with people. Our pluralistic society needs to enjoy urban quality in its many forms, and scenes such as these will continue to provide a lasting contribution to our enjoyment of life.

Sources: Photos of (left) Bristol hilltop view: Gordon Power; and (below) Sarlat (France) crowd scene: Michael Parfect

SARLAT

are to have any real chance of getting to grips with the hugely important but equally diverse problems of ensuring good urban quality.

In this book we have tried to demonstrate, in particular, how the democratic statutory planning system in Britain alternatively aids, enables, hinders (or in some cases serves to discourage) the creation, retention and maintenance of urban quality in this country – with so many lessons for others. From all this, it is evident that, in parallel with the need on all sides for a "proper understanding" of the governing political concept and statutory systems (plus the financial, legal and social forces which constantly underlie and shape our environment), there is an equally important need for both the ordinary citizen and the (otherwise educated) professional simply to *use their eyes* and to take in the evidence that is all around them. Only by studying the physical and visual evidence around us do we acquire the true 'language of environment', and thus hopefully a common ethic, as the first and most essential step towards grasping the realities of urban quality and the problems of its realisation for society as a whole. By this means we take the next step, i.e towards agreement on needs, goals and objectives; from this we may proceed to shape the machinery (of all kinds) that will assist us in achieving the 'end product' of urban quality.

However, if on the other hand we allow urban decay to drift unchecked, the time could come, perhaps not so far distant (as Michael Middleton says) – when the investment required to check it will seem so crippling as to be politically unacceptable. It is a common responsibility, extending right across society, in which we have to discard apathy and detachment, and also put into second place material considerations of personal wealth, comfort, mobility and convenience where they stand in the way of obtaining a better public environment and a better future. In particular, the tyranny of the motor car must be overcome by revised priorities as to transportation, and via modern environmental techniques now appearing in the European home of planning experience.

The 'matrix for the modern age' is now *the integrated, pluralistic urban environment*, and it demands the creation of corresponding statutory systems, development machinery, and the necessary fiscal resources, properly deployed, to provide it. *Ultimately, however, our towns and cities, our environment and our urban quality are quite simply ourselves.* If therefore this book has conveyed to the ordinary citizen as well as to the student of such matters a wider appreciation or understanding of the issues involved, if only as a starting-point, it will have served its purpose.

PART III – SUMMARY OF KEY ASPECTS

A WAY FORWARD: INVESTMENT IN URBAN QUALITY

Socio-economic factors and characteristics, and the role of market forces, form a recurrent thread in the discussion, and are taken further via the need for 'Investment in Urban Quality' ('A Way Forward').

- The next section ('New Directions') brings us back to the need to reinstate the role of local government and town planning (both now at a crossroads) in managing our towns and enhancing their environment.
- Commercialisation of democratic local processes is no answer.
- From our present uncertain position, it is still necessary to look hopefully into an as yet unclear future, rather than retire into the values and forms of the past.
- 'The art of the possible' must continue to be our theme.
- 'The Embodiment of Meaning' both in and of urban quality provides the essential psychological basis for the understanding of the term as studied in our book.

PHILOSOPHY OF CONTEXT

The fundamental importance of context, whether natural or built, forms the theme of this next section.

- The recognition of identity of local urban context, in whatever form, is seen as the essential basis for designing and providing development that can take its place satisfactorily within the urban scene.
- This does not, however, imply slavish conformity to past styles or forms of building, suppressing individuality. Historical evolution is, after all, the process by which identity of place is derived. Numerous examples of philosophy of context exist, and several of these are referred to.
- Quality of proposal relative to quality of context is integral with quality of environment.
- Buildings employing modern technology in appearance and form often pose problems of compatibility, and tend to be more suited to 'stand-alone' locations where they better express individuality, though there will of course be specific exceptions to this.
- The insidious effects of traffic on urban quality are singled out for mention and for consideration.

DEVELOPMENT QUALITY: COMPONENT FACTORS

This short but important section is included in order to create a better understanding of the private developer's viewpoint, via an explanation of the factors he or she must consider when investing financial resources in any development scheme (apart from that of location).

- These factors may be summarised as: finance, amenity, traffic/parking, security/safety and order/organisation.
- As ingredients for successful new development, they clearly contribute in their own way to achievement of urban quality as a whole, whilst not being fully synonymous with it.

THE PUBLIC REALM: NEGLECT, DECAY AND REVITALISATION

Again, a short but pivotal section is included in order to give essential background and understanding of what has been happening to the public realm in Britain from the 1950s onwards. A tabulation is provided, setting out the changing relationship over the last 40 years between prevailing circumstances in the UK and their implications for urban quality in the public environment. This is particularly topical, in view of the correlation between neglect/decay of the public realm and the increased recognition of the legitimate role of the public sector by central government.

THE PUBLIC REALM: A HISTORIC BASIS FOR MODERN URBAN DESIGN/QUALITY

A meaningful study of urban design, however selective or brief, must widen the scope of the term to include underlying economic and social factors that are integral to the organic/dynamic context of the way in which urban areas change and grow.

- The skill of the urban designer must embrace these factors completely, or the end result will be cosmetic, with a limited validity.
- The most successful town planning schemes of the past have been founded upon this truth.

As an introduction therefore to the subject of urban design in modern times we commence with a broad review of the socio-economic investment imperative essential to achieving urban quality on a national scale. Central government in Britain has failed to ensure the necessary supply of fiscal energy on an evenly spread basis, to produce the required improvement in our urban environments.

The authors thus propose that a regional basis of development funding, linked with strategic planning, would be a more satisfactory and effective way of achieving urban quality, via localised urban design, throughout. In short, we need to reinstate the type of 'provincial' approach to the quality of life and its built environment that has so often met with conspicuously successful results in earlier times.

We confine our review of past and present urban design first to a discussion of the qualities/virtues of planned and unplanned town environments.

- Traditionally, urban evolution has taken the shape of either formal or informal layout, frequently a mixture of the two.
- Conscious planning of towns rarely now implies formalised layout on a grand scale, and certainly the British genius (such as it is) has been very much towards informality, with an element of 'fun/muddle'.
- For this to be successful rather than chaotic, however, a thorough understanding of local socio/economic forces plus considerable art and artifice are necessary.
- We have to promote/nurture these skills in producing cohesive, attractive and generally 'user-friendly' environments which have 'style' and 'ambience' in good measure.

URBAN QUALITY: CASE STUDIES[112]

Before looking at various examples of urban quality, we first have to consider a relevant approach or methodology to apply to a range of typical urban situations.

- From the visual evidence, we may derive a number of indicators of urban quality.
- For an in-depth planning assessment of each urban area in practice, however, a more comprehensive analysis would be obtained by considering 'relevance for change' and 'capacity for change', since it is effect of change on urban environments that is our main concern. We look at these questions in turn.
- Under relevance for change, typical urban situations are considered as to likelihood of physical change.
- Then under capacity for change we introduce the concept of environmental capacity, with that of 'environmental capital'. A 1993 BDP/Arup study of Chester is drawn upon to provide relevant first-hand evidence of the applicability and value of this approach.
- Next, indicators for urban quality are looked at, relative to issues for environmental capacity, again using the Chester study as a convenient basis.
- The dual aspects of opportunity and restraint are also highlighted as relevant factors which must be kept in balance at all times.

The fallacy of *laissez-faire* development policies as a 'valid' planning tool, linked to the free play of market forces, is finally mentioned, and reference is made to worldwide failure of this.

MODERN ARCHITECTURE: DESIGN AND TECHNOLOGY

A basic understanding of the 'design process' is needed before a discussion of architecture, whether past or present, can usefully take place.

- This must include an appreciation of matters such as site selection (where any option exists), site planning and floorspace quantum, vehicular and pedestrian access, technological and functional considerations, building and other development costs, and required level of economic return against life/use expectancy for the development.
- All this must be set within relevant urban context, including planning restraints and urban quality hopes and expectations.
- Architectural form and appearance is very much a result or outcome of these many different influences – a fact rarely if at all appreciated by the critical layman, particularly when faced by buildings of modern functional design.

The remainder of this section is considered collectively under the heading 'The Great Divide: Old versus New', as it is clear in many ways that this dichotomy is the ongoing preoccupation of most people in Britain with any interest or involvement in planning and development. This theme often takes precedence over, or runs parallel to, more fundamental considerations.

Subjects discussed under this heading include:-

- new buildings in old/established town contexts;
- focus on styles instead of more relevant factors;
- fixation with the past: Royal connotations, etc;
- the case for innovation and progress.

INTO EUROPE AND THE 21ST CENTURY

The section concludes with an examination of a European planning system – the German one – which has proved a more stable and reliable basis for the production of a consistently high urban quality level than our own inconsistent 'platform' for architecture/development.

INDICATORS, GENERATORS AND CRITERIA FOR URBAN QUALITY

Material for this section is drawn partly from preceding analytical text and partly from visual examples incorporated with detailed commentary.

- The aim is to derive various criteria or indicators required for urban quality achievement from a range of visual examples representative of the types of situation/problem most frequently encountered in our urban areas today.
- Underlying causes or generators are also valuable to apply, though in some cases these can only be surmised.
- The role of design guidance/control is revisited, with reference to examples of local authority design leaflets from an English council experienced/skilled in that field.
- Reference is also made to British design guides and design briefs.
- Contrasted with this is American practice of control of development via planning briefs/ 'design review' only.

SUSTAINABILITY IN URBAN QUALITY

As to sustainability, 'we have the dangerous situation where the environmental demands of society are running ahead of our capacity to deliver' (Michael Welbank, RTPI President, 1992).

- Central government and the public must now face up to the question of recycling land on a major scale, accompanying urban restructuring.
- The obligations of local authorities in this are underlined by Planning Policy Guidance Notes PPG1 and 12).
- In a 'sustainable world', we must plan to live within our environmental means and reduce our demands on scarce resources.
- There is a new concept of environmental capacity deciding the type or amount of development, or determining what sort of 'sustainable' changes can legitimately be allowed to take place.

An essential consideration in all this field will be the capability of local authorities to take on and maintain a full and proper role thereto – requiring financial resourcing of an adequate nature. We should also look at such mechanisms as Community Development Trusts, for what they can provide, again ensuring that they are adequately resourced.

THE WAY AHEAD IN PLANNING AND DEVELOPMENT

The final section of the book starts with the benefits of private sector finance/expertise for a revitalised public realm environment, within a new planned context and an even-handed/ constructive approach to 'fiscal energy' programming by central government.

Any examination of urban quality will undoubtedly benefit from: -

- drawing upon the best examples of European and North American development and planning practice;
- building upon the best examples and trends in British practice, with divergent lessons for others.

The 'matrix for the modern age' is now the integrated, pluralistic urban environment which is in our own hands to shape in a satisfactory manner.

NOTES

PREFACE

1 See also Part II, under 'Publicity, controversy and education' (pp. 98ff.) and 'Education for Urban Quality' (pp. 119ff.)
2 It seemed then that whilst local authorities had a major role in designing and carrying out their own development, there was not much evidence on their part (or of others) of high standards in urban design or environmental improvement. Now, however, the situation is largely reversed; local authorities (increasingly subject to stringent financial and other restrictions) rarely build anything themselves, but are becoming understandably concerned by the need to play a part in and have some control over, the somewhat freewheeling activities (at least up to the recent 1991 recession) of the private development sector. In this, the public have a healthy concern about the standards of design and, more broadly about the impact of urban change on the environment.

PART I

3 The mass movement of materials has in fact been with us since the introduction of railways in the 19th century. In reality, that era signalled the end of vernacular architecture and townscape harmony.
4 An equal, if not greater, threat to urban quality is that of scale of new development. The harmful effects of this are to be seen almost everywhere. See comments on 'Land Use/Floorspace' (p. 62ff.), and 'High Building Strategies/Building Height Policies', etc. (p. 64ff).
5 It was the lack of invasion of Britain after the Norman Conquest which provided a different urban scene from that of the rest of Europe, where city walls were still necessary much later, proliferating much higher densities, apartment living and so on, with only a few exceptions.
6 The Anglo Saxons and the Danes cared little for the sophistication of city life, preferring market centres to be visited for trade, and comfortable villages in which to live, where the Moot Green probably satisfied most of their political ambitions.
7 Unfortunately, in areas desperate for new development, good urban design advice may be swept aside – the point has been made before.
8 There are some difficult public spending and taxation implications for local authorities as to obtaining more funding for design skills. It may be that this could be linked to a widening of the tax base via taxation of betterment; see Eric Reade's book *British Town and Country Planning*, 1987.
9 A betterment tax, however, could be a source of taxation also for regional government: and infrastructure investment could thus become one of its main responsibilities – see later references to regional government, etc. The whole question of betterment appears fraught with problems, nevertheless, from the appointment of the Uthwatt Committee in 1941 (also covering compensation for planning restrictions) to the ill-fated Community Land Act of 1975, and the Development Land Tax Act of 1976, both flagships of the 1974 Labour government.
10 It should be said that preliminary design negotiations, where they are carried out, are normally intended to avoid the latter situation.

PART II

11 RIBA = Royal Institute of British Architects
   RTPI = Royal Town Planning Institute
12 The Essex Design Guide of the 1970s for residential development was the forerunner of these – though each different part of the country should ideally have its own design guide having regard to the importance of local materials and building forms as well as the avoidance of cheap pastiche and unthinking copy of historical styles unrelated to the locality.
13 It is perhaps the changing technological aspects and implications of new buildings which have the most impact upon their form, and which at the same time most worry and concern the public (and thus the local authorities).
14 One can only say that, while this may hold good for certain such modern masterpieces, it hardly applies to the general range of most frequently encountered building proposals in the average type of situation, from the viewpoint of the general public and its local authority. We need to get down to earth a little more, one suspects.
15 Structure plans include county minerals plans and waste plans, together with transportation issues and provisions, inasmuch as these are county-wide issues and considerations.
16 A particular drawback of the appeals system, however, is that design and 'quality' considerations are thereby subject directly to the range of conditions imposed by the Inspector on any consent, unless agreed beforehand between local authority and applicant – hence the extra value of early negotiations.
17 In fact, little is really known about the commitment of most people to historic environments; little empirical research has been done in this area – see for example Hubbard, 'The Value of Conservation', *Town Planning Review* 64(4) 1993.
18 Some councils have in recent times identified Areas of Special Control or similar, but these are of course additional to statutory listing of buildings as groups.
19 Whilst extensions to existing conservation areas, or new designations, are fairly common, deletions are extremely rare (though not unknown).
20 See note 17.
21 However, there should be some recognition that pedestrianisation can go wrong if handled insensitively or too zealously, leading to dead or sterile environments. (The huge central market-place at Minden, Northern Germany, totally emptied both of traffic and most other forms of life, is a case in point.) See also Tibbalds (1992) *Making People-Friendly Towns*, Chapter 6.
22 See p. 42, 'Conservation and Preservation'.
23 See Part III, pp. 201–2, 'Into Europe and the 21st Century'.
24 Though not always to their entire satisfaction – a common problem with specialist or single-issue bodies!
25 The sociological disasters (not to mention also the constructional failures) of the 1960s tower blocks of flats are a case in point.
26 See previous Section, pp. 52–4.
27 See Part III, pp. 198–200, 'The Great Divide: Old v. New'.
28 Some guidance on choice of designer can be given by local councils to enquirers, in the form of lists of local architects, produced by RIBA Regions for this purpose.
29 On the other hand, Hove is a relatively prosperous town with a dozen conservation areas and a few hundred listed buildings. Bradford (West Yorkshire), for example has 57 conservation areas, several of these in inner-city areas housing some of the country's poorest inhabitants, 3,500 listed buildings and at the time of writing they each had a team of three specialists to take care of their heritage.
30 The greatest threat to urban quality in many of our great cities and major towns has little to do with development pressure, or the ability of local authorities to deal with development. It is a simple fact that many urban areas are so desperate for any development at all, that virtually anything can get built, often regardless of quality.
31 Plot ratio is the amount of gross floorspace in a building divided by its plot area, and may be regarded in effect as a simplified form of Floor Space Index.
32 For a practical and objective study of legislation and procedures applying to typical urban quality problems and situations, the best type of detailed overall work that has come to hand in preparation of this present book is *The Which? Guide to Planning and Conservation*, by John Willman, for The Consumers Association, published by Hodder & Stoughton, 1990.
33 The best basic analysis of the problem remains that of Professor Colin Buchanan, in the 1963 landmark study *Traffic in Towns* (Ministry of

Transport 1963). The underlying theme of this was that all urban areas have a traffic saturation level; and that the higher the volume of traffic, the more extensive or profound the works of both traffic improvement and urban rehabilitation need to be. This was not an invitation to build more and better roads within urban areas generally in order to accommodate increasing amounts of traffic, rather it drew attention to the cause and effect relationship whereby the implications of otherwise desirable courses of traffic movement provision had to be seen in terms of their effect on urban fabric and human environment.

34 The Manchester light railway system, based on the old city centre tramway routes, is an excellent contemporary example. Sheffield also has developed a tramway system for 'urban transit'.
35 See p. 51, 'The Development Process in Context'.
36 This is particularly true of large mill buildings etc. in the Yorks/Lancs area. There should be a much stronger level of commitment from central government for the retention of such buildings, via grant aiding, to encourage new uses for them.
37 Additionally, extra incentives can be applied to individual building schemes in the form of awards, plaques, etc. including local authority design awards/commemorative plaques, RIBA/RTPI design awards, and other awards such as the RICS/Financial Times awards.
38 See p. 55, 'The Development Process in Context', under Committee Presentation/Information.
39 In many circumstances, grant levels are however too low to make any significant difference. Section 77 Grants through English Heritage are a case in point. In many inner-city areas where Section 77 grants operate, the take-up is extremely low because non-conservation standard repairs to unlisted buildings cost just a fraction of that for repairs to correct 'conservation' standard. A good example would be the replacement of sliding sash windows and panelled doors, which is doing a great deal to harm the appearance of many inner-city conservation areas.
40 The great exception to this is the English cathedral close, which is generally much more spacious than its French counterpart, and is also surrounded largely by residential buildings relating originally to the central religious function.
41 See 'Other Measures for Urban Quality', under Traffic Measures and Parking Provision (pp. 70–74).
42 There are no easy answers, particularly for an environmentally conscious Planning Officer, to this type of dilemma (and no medals, either).
43 Widely utilised in France, with clear benefits, though apparently not in Britain.
44 Generally used in Britain, though with drawbacks in terms of impact vibration spreading to adjoining properties (and occupants).
45 See *Architecture in the Twentieth Century*, by Peter Gössel and Gabriele Leuthäuser (1992), pp. 292 and 338.
46 The Prince has followed this up by creating his own Institute of Architecture, to further his aims.
47 The astronomer Patrick Moore (who lives at Selsey, West Sussex) refers to this as 'the aurora Bognor Regis'!
48 Originally shown on BBC1, 28 October 1988.
49 The Urban Design Group as major contributors to the promotion of urban quality generally, also sought to describe principles of good urban design as far back as 1986, in their *Towards A Design Group Manifesto*.
50 See also our earlier notes (pp. 104–5) on RTPI Councillors' School, September 1991, 'Achieving Quality in Urban Design', which distils the recommendations from the final chapter of *Making People-Friendly Towns* into four main suggestions for 'getting it right', from the viewpoint of the community and their elected representatives.
51 Referred to in Part I, Introduction, under 'Historical Review/Setting the Scene', p. 7.
52 See p. 113 on *Making People-Friendly Towns* postscript.
53 See reference under 'Achieving Quality in Urban Design' (p. 104), RTPI councillors School, September 1991.
54 They are included because of their 'grass-roots' context and value, as expressed at a national conference. They raise some cogent issues.
55 See, however, following section, 'Education for Urban Quality' (p. 119).
56 Clearly dependent on concerted action between the London boroughs concerned; adequate transportation infrastructure provision, especially public transport; and adequate financial and other resourcing.
57 For example, on subjects such as Local Agenda 21, or purely in-house matters.
58 See special issue of *Planning Week* (RTPI journal), September 1995, on Careers and Education, for list (pp. 8–14).
59 The University of Liverpool (as list above) ran

MCD, DipCD, and MA courses in Civic Design (cf. Urban Design).
60 The University of Manchester (same list) specialised in Urban Design courses, including an MA course in Urban Design and Regeneration.
61 See *Planning Week* special issue of September 1995 (p. 23).
62 Short courses also have value outside the CPD structure as a refresher course after a break, or in preparation for a career move.
63 In this the UK are well behind other European countries, for example the Netherlands, where history of architecture is taught to schoolchildren, giving them a much greater understanding of the development of urban form than their British counterparts have.
64 Unhappily, the Lewis Cohen Urban Studies Centre closed down in the mid-1990s, due to lack of funding.

PART III

65 c.f. Germany, where a longer-term view is taken in business ventures generally.
66 GIS (Geographical Information System) techniques currently being established in the UK in the late 1990s should be of great benefit in this.
67 There is an old adage, however, among terrestrial navigators: 'Man is never lost, only is he uncertain of position.' In the development of his social habitat man has been uncertain of his position, both in space and time, on many occasions throughout his short history on earth. It is indeed possible that the breathtaking development of what we term 'civilisation' over the last 300 years or so, has led to an 'uncertainty of position' in our common attitudes towards our environment, both natural and urban.
68 The now outdated concept of single-use zoning has had a singularly deadening effect, and has now given way to the mixed-uses concept; cf. however the need for defined activity areas, e.g, Central Business Districts, etc.
69 Experiments in new ways of achieving this objective have been under way for some years with Urban Design Workshop initiatives and 'Planning for Real'.
70 Planners are often accused of being obsessed with harmony and 'fitting in', to the exclusion of all else.
71 Had the winning scheme been 'for real', and the subject of a planning application, there were doubts locally as to whether it would have been given formal consent, other than perhaps on appeal. In a different location, however, i.e. with the site not forming part of a historic riverside scene, it is less likely that such opposition would have arisen.
72 cf. Tibbalds' more 'global' criteria for quality in the urban context in general, as expressed in his *Making People-Friendly Towns*.
73 There is, at the same time, increasing concern from many urban designers as to the ways that contemporary places of leisure (shopping centres in particular) are being managed to control or modify user behaviour and attitudes – i.e. the issue of the increasing privatisation of the public realm and the implications for social control.
74 A large amount of the Castle Mall shopping development in Norwich is built either fully or partly below ground level, yet gaining in 1994 (like the Broadgate scheme in the City of London in 1992) an RTPI Silver Jubilee planning and design award.
75 At last, in 1996, Newhaven stood to benefit via a grant from the Single Regeneration Budget, to bring fresh economic life to the town and its port, improving the latter in line with Dieppe.
76 See notes on Security aspects of Development Quality, under Part III, p. 147.
77 This is exemplified by the 1994 DoE discussion document *Quality in Town and Country*, in which unfortunately the visual examples given were very largely of developments funded by national or large-scale public company resources, with little or no reference to the equally basic and important question of the funding required for local-scale development or urban improvements.
78 Many of the industrial towns and cities of the North of England and the Midlands have seen their industrial base heavily eroded in the world recession of the late 1980s and into the 1990s, and furthermore having to compete with Far Eastern economies, by finding new roles and new development at almost any environmental price.
79 Other developments, such as Hampstead Garden Suburb in England, and Radburn, New Jersey in the USA ended up more literally as leafy suburbs than as true Garden Cities.
80 Surprising levels of unemployment, plus social deprivation, are to be found in areas of Southeast England, as well as in the northern half of Britain and in Wales. Seaside towns such as Brighton and Hastings are but two examples.

81 Norwich succeeded where Fort Worth failed because the latter was too ambitious and radical; the British approach was to succeed instead via a set of adaptive measures, skilfully integrated into an overall urban package.
82 Now part of LDR International, whose collaboration on this present book is warmly acknowledged.
83 See also *Planning Week* (RTPI Journal), 1 June 1995, pp. 16,17, for 'Scope for Growth' article on BDPs Chester environmental capacity study (jointly with Arup Economics and Planning). Note also English Heritage's 'Environmental Capacity' report, available from PO Box 229, Northampton NN6 9RY.
84 *Sources for this section*: RIBA *Journal*, April 1993, pp. 21, 29; RIBA *Journal*, September 1995, pp. 30, 31.
85 See again under 'Development Quality', p. 147 (The reference is worth repeating).
86 Produced by students and staff of Brighton Polytechnic (now the University of Brighton) with support from the John and Heather Lomax Student Prize and the South-East Region of the Royal Institute of British Architects.
87 LDR: Land Design Research International of Columbia, Maryland, USA whose generous assistance in numerous respects with the creation and production of this book the authors gratefully acknowledge.
88 For a further appreciation of some of the quite outstanding urban design work carried out in the centre of Birmingham in recent years, see *Urban Design* journal, 54, April 1995, re Victoria Square, for which the City Centre Design Strategy was prepared by Francis Tibbalds' multi-professional consultancy team (article by Geoff Wright and John Blakemore).
89 Sources: 'Sheffield Supertram' article by Andy Topley and Dick Owen-Smith from *Urban Design* no. 54, April 1995 (Urban Design Group Quarterly Journal).
90 See also 'Edinburgh', pp. 140–43.
91 By Land Design Research International (LDR) Columbia, Maryland, USA, jointly with Merseyside Development Corporation.
92 GA Life building, York – see Fig. 30.
93 One of the authors visited the new South Africa in November 1995 – the present report of his reflects the above approach, via his study tour of the Cape Town municipal area, and his discussions with the city's Chief Planner (Policy and Research), Mr Steve Boshoff, whose help is gratefully acknowledged, as is also the help of Geoff and Merle Price generally.
94 President of the Royal Town Planning Institute, 1996/7. See his 'Planning at the Crossroads' article on South Africa in *Planning Week* 14 December 1995, p. 11.
95 Acknowledgements to Barrie P. Watson, Chief Town Planner, City Engineer's Dept, 80, Independence Avenue, Windhoek, Namibia.
96 In particular, we would mention *Architecture in the Twentieth Century* by Gössel and Leuthäuser, Taschen Verlag, Cologne, 1991.
97 Acknowledgements are made to Colin Gosden of Hojgaard and Schultz, international consulting civil engineers and contractors, Charlottenlund, Denmark, for his generous assistance with information on modern Danish projects.
98 In most other mainland European countries also, the planning control and design guidance system is more akin to the German rather than the British model.
99 Information and notes derived from a 1993 paper 'The German Planning System – A Model for Europe?', by John Silvester, Director of Planning and Community Services for Surrey Heath Borough Council, Camberley, Surrey.
100 The illustrations in this book, and their captioning and commentaries, have been formulated so as to provide the opportunity for 'indicators and criteria' for urban quality/design to be derived from or applied to these and similar examples.
101 See *Designing the Successful Downtown* by Cyril B. Paumier (1988).
102 Brundtland Report on world environmental sustainability.
103 John Gummer, UK Secretary of State for the Environment.
104 *Tramway/Light Urban Railway Systems*. Growing popularity for these is to be seen in British northern towns and cities again as at Manchester and Sheffield, after original popular use following Industrial Revolution declined after the introduction of the motor bus/cars. Now, large-scale introduction and revamping of urban light railways is evident across Europe (see Lille, Grenoble, Amsterdam, Prague, etc.). Taxi links are made to principal stops and termini. Concessionary and simplified fare structures for inner-city journeys lead to greater encouragement of public transport, and consequent reduction of private car use.
105 *The 19th century saw the introduction in Britain of*

sound housing and street corner convenience shopping for pedestrian use, close to workplace. It also saw very acceptable low-rent solutions in affordable, hygienic family accommodation, for all age groups. Housing in English northern towns, because of quality in traditional construction, convenience and cheapness, both in sustainability and energy consumption costs, is still in use, where remaining after clearances, now that dirty industries have largely disappeared from urban centres. Close supervision of young children in traffic-free 'play' streets immediately adjacent to homes is provided automatically.

106 *The 19th-century industrial wealth in north Britain produced many high-quality public buildings and public parks at town and city cores*, echoed in most of Britain's provincial industrialised cities (e.g. Cardiff City Centre). The new movement to regain urban quality will need to preserve and enhance such focus and 'monumentality' at high public amenity city centres, in addition to reinstating shopping and providing general 'livability' within pedestrian reach of such public advantages, in up-to-date terms.

107 Development profits in the British house building industry, along the lines which we have discussed immediately above, are analysed in Chapter 3 of *Housing, States and Localities* (Dickens, Duncan, Goodwin and Gray 1985).

108 For a more detailed analysis of these problems, see *The Politics of Mass Housing in Britain, 1945–1975* (Dunleavy 1981). Good design in council housing, however, is dealt with praise in *Homes Fit for Heroes* (Swenarton 1981), describing the efforts to achieve quality.

109 For an excellent overall study of modern trends in UK late-20th-century building design, see *The State of British Architecture* (Lyall 1980).

110 Two other important steps forward in this respect are the Town and Country Planning Association's 1993 study report on sustainability *Planning for a Sustainable Environment*, featuring the development of ideas and mechanisms for a new kind of environmental planning; and the Manchester 2000 Plan, along the same lines also.

111 Anglia Polytechnic University, Chelmsford, Essex, has expertise in this field of technology as does the University of Brighton/Sussex, *inter alia*.

112 See visual examples included.

BIBLIOGRAPHY AND REFERENCES

PART I: INTRODUCTION

BDP Planning, *London's Urban Environment: Planning for Quality*, London, HMSO, 1996.
Cresswell, R., *Quality in Urban Planning and Design*, London, Newnes-Butterworths, 1979.
Cullen, G., *Townscape*, London, Architectural Press, 1961.
Davies, C., *Traffic in Townscape: Ideas from Europe*, London, Civic Trust, 1994.
Department of the Environment, *Quality in Town and Country: Urban Design Campaign*, London, Department of the Environment, 1995.
Gibberd, F., *Town Design*, London, Architectural Press, 1959.
Girouard, M., *Cities and People*, London, Guild, 1985.
Hall, P., *Urban and Regional Planning*, London, Routledge, 1992.
Hiorns, F., *Town-Building in History*, London, Harrap, 1956.
Hoskins, W.G., *The Making of the English Landscape*, London, Penguin Group, 1991.
Hough, M., *Cities and Natural Processes*, London, Routledge, 1995.
HRH Prince of Wales, *A Vision of Britain: A Personal View of Architecture*, London, Doubleday, 1989.
Jones, E., and Van Zandt, E., *The City – Yesterday, Today, and Tomorrow*, London, Aldus Books, 1974.
Lloyd, D.W., *The Making of English Towns*, London, Gollancz, 1984.
Martin, G., *The Town* (in the series *A Visual History of Modern Britain*, ed. Prof. J. Simmons) London, Readers Union/Studio Vista, 1965.
Middleton, M., *Man Made the Town*, London, Bodley Head/Shell UK, 1987.
Movement and Urban Quality in London Symposium, *Movement and Urban Quality in London: Conserving the Public Realm*, London, Vision for London, 1995.
Mumford, L., *The City in History*, London, Penguin/Pelican, 1961.
Plate, K., *From EGO-Mobility to ECO-Mobility*, Dusseldorf, Gemini, 1994.
Reade, E., *British Town and Country Planning*, Milton Keynes, Open University Press, 1987.
Thornley, A., *Urban Planning under Thatcherism*, London, Routledge, 1993.
Whitelegg, J., *Roads, Jobs and the Economy*, London, Greenpeace, 1994.

PART II: TOWN PLANNING FRAMEWORK

Alterman, R., and Cars, G., *Neighbourhood Regeneration – An International Evaluation*, London, Mansell, 1991.
Burke, G., *Town Planning and the Surveyor*, London, Estates Gazette, 1980.
Cormack, P., *Heritage in Danger*, London, New English Library, 1976.
Cullingworth, J.B., and Nadin, V., *Town and Country Planning in Britain*, London, Routledge, 1994.
Galbraith, J.K., *The Culture of Contentment*, New York, Houghton Mifflin Company, 1992.
Glasson, J., Therivel, R., and Chadwick, A., *Introduction to Environmental Impact Assessment*, London, UCL Press, 1994.
Gloag, J., *Men and Buildings*, London, Chantry Publications, 1950.
Gössel, P., and Leuthäuser, G., *Architecture in the Twentieth Century*, Cologne, Benedikt Taschen, 1991.
Guy, C., *The Retail Development Process*, London, Routledge, 1994.

Healy, P., Purdue, M., and Eniss, F., *Negotiating Development: Rationales and Practice for Development Obligations and Planning Gain*, London, Spon, 1995.

HRH. The Prince of Wales, *Vision of Britain*, London, Doubleday, 1989.

Hutchinson, M., *The Prince of Wales – Right or Wrong? An Architect Replies*, London, Faber & Faber, 1989.

Jencks, C., *The Prince, the Architects, and New Wave Monarchy*, London, Academy Editions, 1988.

Keeble, L.B *Principles and Practice of Town and Country Planning*, London, Estates Gazette, 1952.

Keeble, L., *Town Planning Made Plain*, London, Construction Press, 1983.

MacGregor, B., and Ross, A., 'Master or servant? The changing role of the development plan in the British planning system', *Town Planning Review*, 66(1), 41–59, 1995.

Ministry of Transport, *Traffic in Towns*, London, HMSO [Buchanan Report], 1963.

Morris, A.E.J., *History of Urban Form*, Harlow, Longman Scientific and Technical, 1994.

Nairn, I., *Outrage*, London, Architectural Press, 1956.

Nairn, I., *Counterattack*, London, Architectural Press, 1957.

Rasmussen, S.I., *London the Unique City*, London, Penguin/Pelican, 1960.

Reade, E., *British Town and Country Planning*, Milton Keynes, Open University Press, 1987.

Report of Joint Working Group, Institute of Civil Engineers, *Tomorrow's Towns*, London, 1993.

Scottish Office Environment Department *Planning for Crime Prevention*, Planning Advice Note PAN46, London, HMSO, 1994.

Tibbalds, F., 'A Vision of Britain: a review article', *Town Planning Review*, 60(4), 465–67, 1989.

Tibbalds, F., *Making People-Friendly Towns*, Harlow, Essex, Longmans, 1992.

Towers, G., *Building Democracy: Community Architecture and the Inner Cities*, London, UCL Press, 1995.

Planning Legislation Books

Grant, M., *Planning Law Handbook*, London, Sweet & Maxwell, 1981 (and later edns).

Telling, A.E., *Planning Law and Procedure*, London, Butterworths, 1982 (and later edns).

Willman, J., *The Which? Guide to Planning and Conservation*, London, Consumers Association, 1990.

PART III: PERSPECTIVE ON URBAN QUALITY

Ball, M., *Housing and Construction: A troubled relationship?*, Bristol, Policy Press, 1996.

Clarke, L., and Wall, C., *Skills and the Construction Process: A Comparative Study of Vocational Training and Quality*, Bristol, Policy Press, 1996.

Cullingworth, J.B., *Town and Country Planning in Britain*, London, George Allen & Unwin, 1976.

Dewar, D., and Uytenbogaardt, R.S., *South African Cities – a Manifesto for Change*, Cape Town, University of Cape Town, 1991.

Department of the Environment, *Quality in Town and Country* (a) Discussion Document and (b) Urban Design Campaign, London, August 1994 and June 1995, respectively.

Dickens, P., Duncan, S., Goodwin, M., and Gray, F., *Housing, States, and Localities*, London, Methuen, 1985.

Dunleavy, P., *The Politics of Mass Housing in Britain*, Oxford, Clarendon Press, 1981.

Gibberd, F., *Town Design*, London, Architectural Press, 1959.

Giedion, S., *Space-Time and Architecture*, London, Cumberlege, Oxford University Press, 1954.

Girardet, H., *The Gaia Atlas of Cities*, London/Stroud, Gaia Books, 1992.

Gössel, P., and Leuthäuser, G., *Architecture in the Twentieth Century*, Cologne, Benedikt Taschen Verlag GmbH, 1991.

Greed, C., *Women and Planning*, London, Routledge, 1994.

(Report of) House of Commons Employment committee, *The Employment Effects of Urban Development Corporations*, London, HM Government, 1988 (re Docklands development boom).

Howard, E., *Garden Cities of Tomorrow*, London, Faber & Faber, 1965.

Hughes, R., *The Shock of the New*, Thames & Hudson, 1991.

Jacobs, J., *The Death and Life of Great American Cities*, London, Penguin/Pelican Books, 1964.

Kidder Smith, G.E., *The New Architecture of Europe*, London, Penguin/Pelican, 1962.

Little, J., *Gender, Planning and the Policy Process*, Oxford, Pergamon, 1994.

Local Government Management Board, *Local Agenda 21 UK: A Framework for Local Sustainability*, London, Local Government Management Board, 1994.

Lyall, S., *The State of British Architecture*, London, Architectural Press, 1980.

Matrix, *Making Space: Women and the Man-made Environment*, London, Pluto, 1984.

Morris, A.E.J., *History of Urban Form*, Harlow, Longman Scientific and Technical, 1994.

Nasar, J.L. (ed.), *Environmental Aesthetics: Theory, Research and Applications*, Cambridge, Cambridge University Press, 1992.

Norberg-Schulz, C., *Meaning in Western Architecture*, London, Studio Vista, 1975.

Ove Arup and Partners, *Environmental Capacity: A Methodology for Historic Cities*, Manchester, Ove Arup and Partners (for Cheshire County Council, Chester City Council, Department of the Environment and English Heritage).

Paumier, C.B./LDR (Land Design Research International, Columbia, Maryland, USA). *Designing the Successful Downtown*, LDR/Urban Land Institute, 1988.

Price, F.G., CUTA (Committee of Urban Transport Authorities) Land Use-Transport Technical Committee, *Task No. 9: A Review of Instruments of Land Development Policy*, CSIR, Stellenbosch, South Africa, January 1992.

Swenarton, M., *Homes Fit for Heroes*, London, Heinemann, 1981.

Thomas, H., and Krishnarayan, V. (eds), *Race Equality and Planning: Policies and Procedures*, Aldershot, Avebury, 1994.

(Report of) Town and Country Planning Association, *Planning for a Sustainable Environment*, London, Earthscan, 1993.

Tugnutt, A., and Robertson, M., *Making Townscape*, London, Mitchell/Batsford, 1987.

INDEX

Note: Page references in **bold** type indicate illustrations.

Adam, Robert 69, 157
advertisements *see* signs and advertisements
aesthetics 108; architects' freedom 51; controversies 101–3, 127–8; Design Accord between RIBA and RTPI 115–18; failure of post-war projects 148–9, 150; hierarchy, scale and harmony 64–9, 106–7, 126–7, 145, 165; multi-culturalism 222; Nairn's judgement 113; new buildings in historic context 103–4; not a DoE priority 56; *see also* urban quality
Ahrends, Burton and Koralek 101–2
Alsop, Will 192, 194
Alterman, R.: *Neighbourhood Regeneration* (with Cars) 114
amenity societies 54; conservation and preservation 50; educating 120
American Express building, Brighton (Amex House) 66, 67, 147, 173
archaeology: PPG principles 34
architects: freedom 51, 59–60, 116–17; major London site controversies 101–3; perception of quality 98–9; recent controversies 127–8; small percentage of applicants 23–4; Tibbalds' recommendations 113; *see also* modern architecture
Areas of Great Landscape Value 62
Areas of Outstanding Natural Beauty 34
art 108; per cent for 89–90

Barcelona 17, 157
Bath 12, 14, 139–40; Green Park Station 96
Bauhaus School of Design 122
Berlin, war-damaged areas 158
Binney, Marcus: *Preservation Pays* (with Hanna) 48
Birmingham 174, 176–8; Bull Ring development 149
Black, Misha 90
Blake, William 14
Bradshaw, Martin 219
Brighton 140, 173; scale of buildings 65, 66, 67; shopping precinct (Churchill Square) 66, 93
Bristol 167, 223; conservation 43, 47, 172; context 138, 139, 139–40; shopping arcades 97
Britain: building industry 216–17; history of urban development 12–19; *see also* Department of the Environment; government (central)
Britain's Historic Buildings (British Tourist Authority) 87
British Architecture Foundation 123
British Tourist Authority: *Britain's Historic Buildings* 87
Brown, Capability 31
Brunel, Isambard Kingdom 140
Buchanan Report: *Traffic in Towns* 149
building industry 216–17; new technology 190, 192, 194; technology 215–16
Building Regulations: health and safety controls cf. planning applications 53
Burton, Tony 218
business/retail trade: car-oriented shopping 160–1; developmental quality 145, 146; Hove's advice for shopfronts in conservation areas 61–2; lack of design thought for commercial buildings 86; land use/floor space 62–4; marketplace pedestrianisation 92–8; out-of-town development 93, 134, 160; phasing control 74–5; PPG principles 33–4; shopping arcades 96–7; stultifying standardization 154; *see also* tourism
Byker Wall 58

Canterbury 140
Cape Town 179, 181–2, **187–8**, 189
caravans 81–2
Cardiff, dockside/waterfront 179; regional identity/quality 154
Carnwath (Robert) Report 81–2
cars *see* traffic management
Cars, G.: *Neighbourhood Regeneration* (with Alterman) 114
castles 14
Cedar Rapids, Iowa **206–7**
Cerda, Ildefons 17, 18, 157
Ceske Budejovice 158
Charles, Prince of Wales 114; community architecture 58; design guidelines 51, 104, 105–10; Institute of Architecture 115; new versus old debate 31–2, 45, 101–3, 123; stylistic approach 116; support for planners 117–18; Tibbalds' response to 111–12; *A Vision of Britain* 2, 20, 100, 128

INDEX

Chester 12, 163, 165, **166, 167**, 168, 169, 227
Chichester 140, 155
churches 14; cathedrals as urban centres 140; spires and towers sit well visually 65–6
Civic Amenities Act (1967) 42
Civic Trust 91–2, 219
Colchester, Lion Walk precinct 93
Collinson, Harry 82
communities 57–8, 109; civic pride 85; *see also* local style
Community Development Trusts 219–20
Comprehensive Development Areas (CDAs) 37
conservation and preservation 42–3, 125–6; capacity for change 160, 165; case studies 172; commercial uses 88; compulsory purchase 48; criminalisation of offences 82; demolition 75–6; financial resources 45, 92; infrastructure 89; issues and indicators of quality 168–9; land use 46; obligations of owners 46, 48; perception of quality 45–6, 98–9; planning applications 40; power of local authorities 43–5; PPG principles 34; preservation pays 48–9, 85, 87–8, 127; problems with historic buildings 59; railway stations 71; sympathetic continuation 31, 103–4, 156–8, 200; town's loss of role 161, 163; unlisted buildings 49–50; way to proceed 50–1
consultants: planning applications 40
context 161; Bristol 172; Edinburgh 172; La Rochelle 169–71; Lewes 173; modern architecture 227; modern pluralism 201; philosophy of 137–45, 225
Copenhagen International Airport 190, **191**, **192**
Le Corbusier (Charles Jeanneret) 158
Counter-Attack (Nairn) 8, 113
Craig, James 141, 172
crime: PPG priniciples 33; preservation offences 81–2; security in development 147
Cullen, Gordon: *Townscape* 2, 7, 8, 100, 113–14, 128
The Culture of Contentment (Galbraith) 83

Danish building excellence 190, 192, **193**
Defoe, Daniel 175
demolition 61, 70, 75, 160
Department of the Environment: appeals to 39, 41, 55–6; design guidance 22–3; grant aids 91; listed buildings 44; local authority's role 30; quality campaign 118–19; *Quality in Town and Country* 136, 152–3, 212; *Quality of Life in Cities* 154–5; Secretary of State on design guidelines 38–9
Department of Transport 41
derelict/vacant sites 70, 76–7
design advice and control 25–6; advice resources 29–30; aesthetic freedom 51, 59–60; codes and themes 205; context 125, 161; government guidance 22–4; methodology towards urban quality 204–6; model guidelines 51; new versus old 30–2; Prince Charles's 10 principles 105–10; quality from different points of view 98–9; societies and associations 54; Tibbalds' suggestions for getting it right 104–5
Design Guidance Codes 38
Design Guide for Residential Areas (Dunbar and Essex County Council) 86
Designing the Successful Downtown (Paumier) 161

developers: versus architects 23–4
development plan system 35–7
Dublin 11, 104
Dunbar, Melvin: *Design Guide for Residential Areas* 86

economics: benefits of preservation 87–8; changing attitudes about public and private sectors 83–5; developmental quality finance and maintenance 145–6; educating financing institutions 120; enhancing property and land value 85–7; grants 90–12; investment 133–4, 225; *laissez-faire* development 69–70; market forces not sensitive to social context 151; modern architecture 198, 200–1; result of drift into urban decay 224; socio-economic factors 150, 152–4
Edinburgh 140, **141**, **142**, **143**; context and conservation 172; International Conference Centre 172
education 127–9; for designers 121–3; differences in outlook 198, 200; environmental 123–4; learning about quality from history 136; public awareness 119–21
enclosure 107
English Heritage 40, 54, 92
English Historic Towns Forum 49, 125–6
English Partnerships (Urban Regeneration Agency) 91, 210
enhancement opportunities 48, 49
Enterprise Zones 36
environment: balance sheet for quality 203–4; capacity 163, 165, 168; context of development proposals 51–2; education 123–4; effect of Green movement 114; issues and indicators of quality 168–9; PPG principles 32–3; professional responsibility 219; Rio Summit 212–13; Tibbalds' points 104; *see also* pollution; sustainability; traffic management
Environmental Aesthetics (Nasar) 145
Epsom: Ashley Centre 71, 96, 148; traffic 71
Essex: *Design Guide for Residential Areas* 86
European Union: grants and loans 92
Extensions to the Rear of Properties (Hove Borough Council) 60–1
Eyre, Sir Graham 46

facelift schemes 85, 90
Farrell, Terry 172, 192, 201
Fidler, Peter 118
floor space *see* land use/floor space
Foot Streets in Four Cities (Wood) 96
Fort Worth, Texas 161, **162**, 174
Foster, Norman 192, 201
Fyson, Anthony 155

Galbraith, John Kenneth 127; *The Culture of Contentment* 83
Garden Cities of Tomorrow (Howard) 158
Georgian Group 54
Germany: building/design quality 190; planning system 201–2; South German 'village towns' 155
Ghent 179, **186**
Gibberd, Frederick: *Town Design* 2
Gibson, Sir Donald 96

Glasgow 158, 175–6, 179, **183**
global warming 33
Gollins Melvin and Ward 66, 173
government (central): clear advice 113; long-term considerations 113; squeezes local resources 22; won't/can't pay 153; *see also* Department of the Environment
grants 90–2
Green Belts 62, 63
Greenville, South Carolina **164**, 174
Gropius, Walter 122, 158
Gummer, John 119, 155

Hague, Cliff 189
Hanna, Max: *Preservation Pays* (with Binney) 48
Hastings Trust 220
Haussmann, Baron Eugène Georges 18, 157
Heseltine, Michael 38–9
hierarchy 106
high buildings policies 64–9, 126–7
Highways Authorities 71, 72
Hillman, Judy: *Planning for Beauty* 50, 51, 115, 120
Historic Buildings Council for England 87
Hopkins, Michael 201
Hordern, Richard 201
housing: extensions, conversions and alterations 60–1; nineteenth century workers' 14, 18; PPG principles 33; tower blocks 150
Hove 60; *Extensions to the Rear of Properties* 60–1; *Shopfronts in conservation areas* 61–2
Howard, Ebenezer: *Garden Cities of Tomorrow* 158
Hughes, Robert: *The Shock of the New* 199
Hutchinson, Maxwell: *The Prince of Wales – Right or Wrong?* 31–2, 100

industry: growth during Industrial Revolution 14; noise pollution 79; PPG principles 33
infrastructure: master plans 36–7; phasing control 74
Institute of Architecture 115
Institute of Civil Engineers: *Tomorrow's Towns* 118
Investment in urban quality 133–5

Jencks, Charles 59

Keeble, Lewis: Floor Space Index 64; *Principles and Practice of Town and Country Planning* 63
Koblenz, historic skyline 66
Krier, Leon 68, 153

La Rochelle 169, **170–1**
laissez-faire development 69–70
Lakeland, Florida 210–11
land use/floor space 62–4, 126–7; context 137, 139, 161; flexibility 46, 87; Floor Space Index 63–4; husbandry of the land 217–18; mixed use 136, 147–8; order and organisation 145, 147–8; sustainability 214–16
landscapes 89; blending buildings 105–6; urban scene 97–8
Layout in Residential Areas (Surrey) 86–7
Le Havre 37
Letchworth 158

Lewes 65, 72, 140, 144, 173; sympathetic continuation 156–7
Lewis Cohen Urban Studies Centre 123
lifestyle 152–4
lighting 109
Lincoln 13, 47, 140
listed buildings 40; classification 44; demolition 54, 75; local authorities 43–4; obligations of owners 46, 48; planning applications 51; PPG principles 34; *see also* conservation
literature 2; advice leaflets and handbooks 60–1; influential publications 113–15; publicity and pronouncements 100–1; *see also* Charles, Prince of Wales
litter 78–9
Liverpool 179, **185**
local authorities: advice leaflets and handbooks 60–1; compulsory purchase 48; conservation and preservation 43–51; controlling unauthorised development 80–2; design guidance 22–4, 25–6, 120, 204–6; dual aspects of opportunity and restraint 169; economic returns 22; educating 122; global versus detail concerns 30; local plans 35; overall urban design 105; power over urban quality 19–21; processing proposals 39–42; quangos replace duties of 152; routine and small proposals 29; specialist staff 40–1, 113; structure plan policies 34–5
local style: building materials 107–8, 213; county design guides 86–7; DoE guidelines 38; jumbled 10; prescriptions 118; Prince Charles on 105–6; provincial approach 152–4, 226; sense of community 109; stultifying standardization 154; sustainable standards 26
London: Barbican Centre 162, 174; Broadgate 94, 173; Canary Wharf and Docklands 67–8, 93; effect of cars 17; Loughborough Road housing estate **163**, 174; major controversies 69, 101–3; Roman influence 12; skyline 67–8

McCormac, Richard 123
Making People-Friendly Towns (Tibbalds) 100, 112–13, 128
Man Made the Town (Middleton) 21, 25, 222, 223
Manchester 174–5, **180**, **181**, **183**; light railway 175, **182**
Manser, Michael 59–60, 116, 118
Mansion House 101–3
Marseilles 158, 192, 194, **194**, **195**
Michelangelo, design for Capitol, Rome 174
Middleton, Michael: *Man Made the Town* 21, 25, 222, 223
Mies van der Rohe, Ludwig 101, 158
Milton Keynes 158–9
Ministry of Town and Country Planning: *The Redevelopment of Central Areas* 63
modern architecture 227; context and process 197, 201; controversy of old versus new 195, 198–200; design ethic 195, 197; economics and innovation 200–1; focus on style 199; new technology 190, 192, 194, **196**; shock of the new 199
Montpellier, modern civic layout 158

Nairn, Ian: *Counter-Attack* 8, 113; *Outrage* 113
Nancy, Place Stanislas 158
Nasar, J.L.: *Environmental Aesthetics* 145

National Gallery 101–2
National Heritage Memorial Fund 92
Neighbourhood Regeneration (Alterman and Cars) 114
new versus old 30–2, 103–4, 123; controversy in modern architecture 195, 198–200; historical roots of design 155–8; modern architecture 227; new technology 190, 192, 194, **196**; new towns 158–9; philosophy of context 144
Newcastle: waterfront 179, **184**
NIMBY (Not In My Backyard) 45–6, 69–70
noise pollution: PPG principles 33
Norberg-Schulz, C. 136
North Kensington Amenity Trust 220
Norwich 95; Castle Mall 148, 161, 173; London Street dual pedestrian precinct 49, 90, 93; organisation in mixed use 148; ring and loop system 96, 161

offences, controlling 82
organic cities, towns, villages 155, 156
Outrage (Nairn) 113
Oxford Brookes University 121
Oxford, St Ebbe's 63

Palumbo, Peter 102
Paris: Haussmann redesigns 18, 157; La Défense 93; Pompidou Centre 137
St Paul's Cathedral 68, 69, 107
Paumier, Cyril B.: *Designing the Successful Downtown* 161
pedestrianisation: enhancement opportunities 48–9; *see also* traffic management
Percent for Art (Arts Council) 89–90
Perspectives (journal) 115, 120
phasing control 70, 74–5
planners: education 121–3
planning: briefs 37–9, 205; city/town centres 160; Comprehensive Development Areas (CDAs) 37; criteria for quality 203–4; flexibility 134–5; gain 49, 89, 146; historical roots 155–8; humanisation and use of theme 204; infill development 160; major schemes 58–9, 157; master plans 36–7; merits and demerits 23; new towns 158–9; opportunity versus restraint 21–2, 169; optimistic ideas of 1960s 84; perception of quality 98; planned evolution/sympathetic continuation 156–8; public participation 56–8; relevance and capacity for change 159–61; result of drift into urban decay 224; sensitive areas 58–9; site plan review 205; unauthorised development 70, 80–2
Planning Act (1971) 42
planning applications: appeals 55–6; the applicants 39–40; architects' 23–4; British flexibility 202; development process in context 51–4; extra concessions by developers 56; extra fees for commercial proposals 51; German strictness in procedure 201–2; objections 40, 54–5; pre-application negotiations 125; presentation to committee 55; process 53; process of proposals 39–42; summary of process 126
Planning for Beauty (Royal Fine Art Commission/Hillman) 50, 51, 115, 120
Planning for Women (RTPI) 221
planning gain 49, 89, 146

Planning (Listed Buildings and Conservation Areas) Act (1990) 43–4, 45–6
Planning Policy Guidance Notes (PPGs) 23, 115; economic activities 33–4; environmental principles 32–3, 218; historic environment 34; plan and control systems 32; transport and communications 34
politics 99; policy formation 56–7
pollution: noise 79–80; PPG priniciples 33; traffic 10, 79–80, 144; *see also* traffic management
Poundbury 153
Prague, historic skyline 65, 67
Preservation Pays (Binney and Hanna) 48
press and media: influence on public opinion 99–100
Priene, Asia Minor 155
The Prince of Wales – Right or Wrong? (Hutchinson) 31–2, 100
Principles and Practice of Town and Country Planning (Keeble) 63
private and public sectors 127, 174–5; cooperation for quality 206–7, 210, 212; potential for joint action 83–5; street facelift schemes 90
property and land value: increasing 85–7; property now dominant 135
public areas 226; Birmingham's civic renovation 174; failure of post-war projects 148–9, 150; roads 149–50
public opinion: and aethetic choices 59–60; conservation 45–6; consultation and participation in planning 56–8; design control 24; growing awareness of issues 114–15; influence 1, 2; objections to planning applications 54–5; perception of quality 99; press and media influence on 99–100; publications 100–1; publicity and pronouncements 100–1, 127–8
public sector *see* private and public sectors

Quality in Town and Country (DoE) 136, 152–3, 212
Quality of Life in Cities (DoE) 154–5
Queen's University of Belfast 121

racial, ethnic and religious groups 222
The Redevelopment of Central Areas (Ministry of Town and Country Planning) 63
Rees, Peter 116
refuse collection: privatisation 79
residents' associations 54
Resources for Environmental Education (Lewis Cohen Centre) 123
Rio Summit 212–13, 218
Roebuck, Sandra 168
Rogers, Sir Richard 192, 201; new versus old 31–2; sustainability 214
Roman Empire 12; images of Ancient Rome/Roman antiquity 14
Rome 15; Baroque city of Sixtus V, etc. 147; St Peters 93
Rotterdam, Lijnbaan shopping precinct 96
Rowland, Jon 219
Royal Fine Art Commission: design control 54; *Planning for Beauty* 50, 51, 115, 120
Royal Institute of British Architects (RIBA) 92, 219; Design Accord with RTPI 113, 115–18
Royal Town Planning Institute (RTPI) 111, 219; Design

Accord with RIBA 113, 115–18; education 121–3; *Planning for Women* 221; Tibbalds' suggestions for getting it right 104–5; summer schools 120
rural areas 8–9

St John of Fawsley, Lord 120
Salisbury 12, 93, 140
Sandys, Duncan 42
Sarlat 223
satellite dishes 61
scale and harmony 64–9, 106–7; *see also* aesthetics
Schinkel, Karl Friedrich 157
security in developmental quality 145, 147
Sheffield, supertram system 175
The Shock of the New (Hughes) 199
Shopfronts in conservation areas (Hove Borough Council) 61–2
signs and advertising 70, 77–9; flyposting 78; Prince Charles's principles 108–9; traffic 10
Simplified Planning Zones 36
Simpson, John 68, 69
Single Regeneration Budget 91
Skeffington Report 56
Smigielski, Konrad 48
Smithson, Peter and Alison 122
Society for the Protection of Ancient Buildings (SPAB) 54
socio-economic factors 150, 152–4; racial, ethnic and religious groups 222
South West Regional Health Authority v. Secretary of State 46
Southfield, Michigan 208–9
Stamp, Gavin 102, 103–4
Stansfield-Smith, Colin 122–3
Steinberg and Sykes v. Secretary of State 45–6
Stirling, Sir James 102
Strasbourg, city centre trams and terminus **215**
subtopia, urban design v. **8**, **9**
superstores 10
Surrey: *Layout in Residential Areas* 86–7
sustainability 228; building technology 215–16; defined 213; husbandry of the land 217–18; Local Agenda 21 212–13; mechanisms for quality 218–20; relevance to urban quality 213–14

taste *see* aesthetics
technology 190, 192, 194; context 225; sustainability 215–16
Terry, Quinlan 69
Tibbalds, Francis 114; fun/muddle 140; *Making People-Friendly Towns* 100, 112–13, 128; suggestions for getting it right 104–5; ten commandments 110–12, 128; tribute to 111
Tibbalds Karski Monro Consultants 64
Tomorrow's Towns (Institute of Civil Engineers) 118
tourism 85; benefit of preservation of old buildings 48–9; commercial use of historic buildings 88; PPG principles 34
Town and Country Planning Act (1947) 11, 17
Town Design (Gibberd) 2
Townscape (Cullen) 2, 7, 8, 100, 113–14, 128

Traffic in Towns (Buchanan Report) 149
traffic management 10, 70–1, 127; assessing planning applications 53–4; cars most damaging influence 140, 144; circulation within town 72, 73–4; conflicts between pedestrians and vehicles 168; conservation areas 49; developers' contribution to roads 146; developmental quality 145, 146–7; environmental deterioration 70–1, 79–80, 127; M25 Greater London orbital 149–50; noise 79–80; parking areas and structures 72–3; pedestrianisation 92–8, 98; PPG highways control 34; provision of safe parking for women 221; signs and lights 109; street parking 71–2, 74; tyranny of the car 7–11, 73, 134, 149, 160–1
transport: bicycles 72, 74; British over-use of cars 149; highways development 34; keeping in character of town 165; light railway systems 175; planning applications 41; PPG principles 33–4; public 74; railways 15
Transport Preparation Pool (TPP) 71
trees 61; facelift schemes 85; Tree Preservation Orders 75, 82

unitary development plans (UDPs) 32, 34–5
United States: design codes 205; LDR projects **164**, **206**, **207**, **208**, **209**, **210**, **211**; 'Seaside', Florida 205
University of Liverpool 121
University of London 121
University of Manchester 122
University of Newcastle 122
Urban Design Campaign 119
Urban Design Group 218–19
urban quality: component factors 145–8, 225–6; defining 135–7; establishing criteria 203–4, 228; issues and indicators 168–9; mechanisms for sustainability and quality 218–20; perception of 98–9; philosophy of context 137–45; renewal 159–60; successful cases 173; sustainability 213–14

Venice 98, 156
Venturi, Robert 101
Victorian engineering 14
Victorian Society 54
Vienna 157; Ringstrasse 157
A Vision of Britain (Charles, Prince of Wales) 2, 20, 100, 128; 10 principles of design 105–10

waterfronts 179, **184**, **185**, **186**; Cape Town 182, **187**, **188**
Welbank, Michael 115, 217
Welwyn Garden City 158
Windhoek 189, **190**, **191**
women 221–2
Wood, Alfred 49, 93, 128; *Foot Streets in Four Cities* 96
Woods of York and Bath 14
Woolley, A.R. 63
Wotton, Sir Henry 31
Wren, Sir Christopher 14, 68
Wright, Frank Lloyd 158

York 12, 14, 140, 179, **186**
Young, Sir George 218